Cross-Cultural Biotechnology

Cross-Cultural Biotechnology

Edited by
Michael C. Brannigan

ROWMAN & LITTLEFIELD PUBLISHERS, INC.
Lanham • Boulder • New York • Toronto • Oxford

ROWMAN & LITTLEFIELD PUBLISHERS, INC.

Published in the United States of America
by Rowman & Littlefield Publishers, Inc.
A wholly owned subsidiary of The Rowman & Littlefield Publishing Group, Inc.
4501 Forbes Boulevard, Suite 200, Lanham, Maryland 20706
www.rowmanlittlefield.com

PO Box 317
Oxford
OX2 9RU, UK

British Library Cataloguing in Publication Information Available

Library of Congress Cataloging-in-Publication Data

Cross-cultural biotechnology / edited by Micheal C. Brannigan.
   p. cm.
  Includes bibliographical references and index.
  ISBN 0-7425-3266-6 (hardcover : alk. paper) — ISBN 0-7425-3267-4
(pbk. : alk. paper)
  1. Biotechnology—Cross-cultural studies. I. Brannigan, Michael C., 1948-
  TP248.23.C76 2004
  660.6—dc22

                                                    2004003350

Printed in the United States of America

∞™ The paper used in this publication meets the minimum requirements of
American National Standard for Information Sciences—Permanence of Paper
for Printed Library Materials, ANSI/NISO Z39.48-1992.

Through these doors no strangers pass,
only friends who have not yet met.
(over an entrance to an Irish pub)

To all who have constructed, cultivated, and contributed to the astounding success and enrichment of the Salzburg Seminar. You continue to inspire with your profound vision of cross-cultural dialogue in one global community. May the world take note of your labor and example.

Contents

Preface and Acknowledgments

The order and pace of events have their way of impressing upon us the immediate tasks at hand. This is so because no actions occur in isolation. All things are strangely interconnected. One sure proof of this lies in biotechnology and its applications. For further evidence, consider the inspiration behind this book.

When I was first invited to participate in a Salzburg Seminar on "Biotechnology: Legal, Ethical, and Social Issues," I did not anticipate that it would turn into one of the most enriching experiences in my career. In October 2001, just over fifty of us representing thirty countries convened in Salzburg, Austria, to discuss various aspects of biotechnology and its far-reaching global implications. This was five weeks after the events of September 11, 2001. On the surface, the contrast between those events and the venue of our conference could not be any more striking, as if we retreated from a cauldron of international tension to a serene sanctuary nestled in the Austrian mountains.

Our conference was held in the idyllic setting of Schloss Leopoldskron (the location for part of the classic film *The Sound of Music*). The mood was restrained yet hopeful. In some respects, biotechnology is the great equalizer, reminding us that at our most basic, molecular level, we are fundamentally the same. And the timing of events no doubt compelled us to be all the more acutely aware of our common human ground. Furthermore, historical necessity painfully reminded us of the moral imperative behind our charge to seriously consider the profound global consequences of biotechnologies. Most importantly, I believe that our collegiality and fellowship reaffirmed for us the need to cultivate genuine cross-cultural dialogue. This dialogue can transform cataclysm into calm, and heartbreak into hope. We gathered together with composure, passion, and faith in our ability to transcend our differences and to reconnect with our common lifeline.

This book investigates biotechnology within a global and cross-cultural context. It poses this critical question: Can we cultivate a shared humane vision concerning biotechnology and its applications? Biotechnology is a two-edged sword. For every gain there is a trade-off. Are the trade-offs worth the investment? Over the long haul, will the world be better off? Will the world's starving be fed? Will the poor remain poor?

Part one takes us on a tour of major global issues and international policies. Here we see the ubiquitous tension between commercialization and equitable access. The need for global partnership is clear. Part two then goes on to examine specific biotechnological challenges within various cultural settings: genetic research in the United States; genetic testing and regulatory concerns in Canada; embryonic research in Europe; the overcoming of past legacies in the former Soviet republic; Jewish and Islamic perspectives on biotechnologies; food security issues in Africa; Confucianism in Asia; and the role of indigenous cultures. Part three explores three global challenges: the need to balance intellectual property rights and fair access; the need for media sensitivity to cultural contexts; finally, the need to better understand and prepare for what may well be inevitable—bioterrorism.

Although Western voices still dominate the discussion when it comes to biotechnologies, this book claims that it is high time to listen to other viewpoints from other cultures. Yet these views embrace a global voice and message: Cross-cultural bridge-building is needed now, more than ever.

My deepest thanks to those on the Rowman & Littlefield staff. Editor Eve DeVaro encouraged and guided me from the start. Her patience, astute advice, and keen support made all the difference. Her able assistant Tessa Fallon steered us through the process in the most sincere and professional manner. Michael Marino skillfully led us through the final stages of production.

Casey Kutcher has done a superb job in organizing the material and working with our authors. In the same fashion, she continues to administer our Institute for Cross-Cultural Ethics with aplomb and genuine commitment. This book bears witness to the mission of our institute. Darlene Veghts has done a remarkable job copyediting. She exudes patience and competence.

I can never thank enough the real authors of this book, those who contributed through their thoughtful, insightful, and enlightening articles. They are all experts in their various fields. Their professionalism shines through in their work. But more importantly, they are each personally committed to excellence as they share the hopeful vision of global dialogue.

My sincere indebtedness to all of the Salzburg Seminar staff who made this encounter possible. They include: Gotz Opitz, Debbie Hooper, Kevin Li Coll, Chivas Dabbs; Salzburg Seminar president Olin Robison; director Tim Ryback; alumni officer Marie-Louise Ryback; librarians Eva Mittendorfer and

Nathalie Phelps; Bernadette Prasser; Robert Jager; program director Katherine McHugh Lichliter; and program officer Helen Maciejewski. A special note of thanks to Charlotte Harrison. I would never have had this enriching experience without her generous confidence and encouragement.

Special tribute goes to every one of the delegates who played an intricate role in our sessions. Some of them have skillfully contributed to this book. All others have contributed in equally valuable ways by their active engagement, questions, astute insights, and rich presence: Hans Galjaard, Johannes Huber, Julian Kinderlerer, Bartha M. Knoppers, Emmanuel Agius, Geetha Bali, Shoshana Berman, Sujatha Byravan, Chantal Line Carpentier, Ricardo Castro Gonzalez, Constance Hon Yee Chan, Serguei Victorovich Chebanov, Michael Chirara, Kalpana Chittaranjan, Jade Donavanik, Marceline Egnin, Jiri Fajkus, Heidi Grasswick, Saturnina Halos, Jan Hartman, Dirk Heumueller, Giles Jackson, Vanya Koleva, Werner Krebs, Sviatlana Kurachkina, Marina Kvatchadze, Maria Fidelis Manalo, Rodrigo Martinez, Reka Matolay, Godwin Mbamalu, Shane Morris, Oleg Novozhylov, Alexander Odaibo, Muriel Poston, Vardit Ravitsky Neumann, John Edward Rege, Margaret Richards, Henry Riggs, James Robeson, Lee Roecker, Maria Sarquis, Anat Scolnicov, Louise Pannenborg-Stutterheim, Mark Taylor, Sylvia Tomova, Jean Charles Van Eeckhaute, Susan Wallace, Stephen Wright, and Zhang Youyun.

As for my dearest friends and family, my gratitude will never be enough. And, as her support knows no bounds, my wife Brooke encourages me without fail.

Introduction: The Need for Cross-Cultural Perspectives in Biotechnology

Michael C. Brannigan

Biotechnology clearly crosses cultural boundaries. In the early 1990s, the National Institutes of Health (NIH) funded a research project to study the DNA material from a remote Panamanian Indian tribe known as the Guaymi. There was reason to believe that the DNA could help produce an antibody for treating both AIDS and leukemia. After a researcher from NIH developed a cell line from a Guaymi Indian woman's blood sample, NIH then sought a patent on the cell line. This generated cross-cultural shock waves and protests that this was outright exploitation of indigenous groups for richer countries' profit. Critics also opposed the idea of patenting peoples. Supporters argued that since the Human Genome Project excluded studies of genetic variations in numerous other cultures, the Human Genome Diversity Project (HGDP) was intended to expand the scope of research to include indigenous peoples. The knowledge culled from this research could be extremely medically useful. When the U.S. biotech company Agracetus was given a patent in 1992 for cotton plants that had been genetically modified, this bestowed wide-ranging ownership to any genetically engineered cotton plants. This in turn enabled the company to obtain patents in other cotton-producing countries such as India and China. At the same time, this placed limits on the abilities of Indian and Chinese scientists to come up with their own genetically engineered cotton plants. What about the increasing global distribution of modified livestock and crop seeds? Many fear that this diminishes worldwide food diversity. Moreover, such distribution increases the likelihood of dependency upon a few multinational companies who hold patents on genetically modified animals and plants. In all this, we continue to witness the far-reaching effects of biotechnological applications upon the global market, global research, and other cultures.

The progress of biotechnology echoes the resounding truth behind Francis Bacon's simple equation in his *Novum Organum*—that *scientia est potentia*, "knowledge is power." And because knowledge can be used for better or for worse, we have a moral imperative to use our knowledge only for the good of life on this planet and beyond. Biotechnology can thereby only make sense if we apply it in ways that alleviate suffering and enhance well-being.

Let us pause and ask: What *is* biotechnology? The term is in common parlance these days, yet if we try to pin it down, it is not that easy to define. Au courant though the term may be, "biotechnology" remains spurious, elusive, and susceptible to a host of misconceptions. Therefore, one principal aim behind this book is to inform and enlighten us as to what biotechnology is as well as its numerous applications.

On the most fundamental level, we at least know that biotechnology has to do with the numerous functions and techniques surrounding cell biology. The human body contains more than one hundred trillion cells: that's a lot of material to work with. As Michael Morgan and Susan Wallace point out in their essay, if stitched together, the length of all the DNA in a single human body could make three round-trips to the sun. In any case, as for a good working definition of biotechnology, I like Martin Makinde's definition of biotechnology as "the manipulation and use of biological organisms to make products that benefit human beings."

Here is a caveat. What biotechnology *is* does *not* rest solely upon how it is depicted. As Margaret Coffey reminds us in her essay near the end of the book, we need to be cautious of media representations of biotechnology that tend to focus upon biotechnology's more stunning potential, such as using genetically modified foods to prevent diseases like cancer or genetically engineering animals to produce effective retroviral agents and drugs. We also hear about bizarre experiments, such as the recent claim (in November 2003) by South Korean researchers that they have "successfully" inserted human embryonic stem cells into the embryos of mice. This compels us to ask, What is the point? Why produce an organism with more than one genetic set of cells, a chimera? How far are we willing to go to produce the "perfect" baby, the "perfect" human?

A sound risk–benefit assessment is critical. For instance, does food from animals that are cloned pose a health risk to consumers? At the start of November 2003, the U.S. Food and Drug Administration (FDA) officially reported that cloned animals are generally safe to eat. Prior to its statement, the FDA had placed a voluntary ban on eating food products from such animals by asking companies practicing animal cloning to refrain from selling food products. Yet questions still abound. Even if the cloned animals are viewed as "healthy," does that ensure that their food products are without risk to consumers? Indeed, this same question can be asked of animals that are not cloned, but are considered to be healthy. Do we have enough data at this point to ensure that food from cloned animals is without health risk? From what we know so far about cloned animals, there appear to be tendencies toward complications such as shorter life expectancy, liver and kidney problems, and hypertension. Aside from safety issues, what about the moral issue of genetically engineering animals for "better" qualities such as more tender meat, or milk with less lactic acid? In weighing the balance of risks and benefits, we must persist in inquiring about biotechnology's impact on the environment, its affect on health care, its repercussions on society, and the relevance of cultural and religious values.

Consider other questions. Should there be limits with respect to accurate labeling of foods, particularly genetically modified foods? What about the public's right to be properly informed about the contents of their food in order to make informed choices? How far should we require genetic screening, especially when certain conditions that are diagnosed do not yet have effective treatments? What about the benefits of preimplantation genetic diagnosis versus its risks? Jeffrey Kahn and Anna Mastroianni give us a discerning discussion of this tension in their essay that thoughtfully examines the Molly Nash case. And what about gene therapy? Although much of somatic gene therapy (affecting the individual patient) seems to have been accepted, there are thornier issues over altering germ-line genes since these affect the future progeny of the patient. While germ-line therapy may well be the only way to prevent hereditary disorders, critics point out how sanctioning this type of therapy is but another step down the slippery slope of abuse. Moreover, although these therapies seem to support individual rights and freedoms, how will they affect privacy issues in addition to insurance coverage for future generations? These questions continue to remind us that we cannot separate what biotechnology *is* from its *goals*. What are the goals of biotechnology? Is it ultimately to benefit the world? If so, then why do the poor nations remain poor? Despite the declared aim of increased food production, will the millions who daily starve actually benefit from the applications of biotechnologies? A glaring example of inequity within our global health-care expenditures rests on what is referred to as the "10/90 gap." This refers to the fact that at least 90 percent of health-care funds worldwide are spent on 10 percent of the world's population.[1] Does this same inequity reveal itself when it comes to applications of biotechnologies? Despite promises of improved healthcare, will those who suffer the most from the lack of basic health services stand to gain?

Perhaps such questions can be better served by framing them in terms of key determinants that measure whether or not biotechnologies are genuinely effective. A recent study singles out six considerations: impact, appropriateness, burden, feasibility, knowledge gap, and indirect benefits.[2] What is the overall *impact* of biotechnology when it comes to improving the health and well-being of peoples? Are biotechnologies generally *accessible* for the population? Are they affordable? This is especially critical in developing countries. For example, will vaccines against HIV be made accessible and affordable for those in African countries who suffer the most from AIDS? As Dianne Nicol astutely asks in her essay, will patenting by major companies lead to easier access and affordability for poorer nations? How *appropriate* are these biotechnologies within specific cultures? Do they conflict with cultural and societal values? Will biotechnologies help to deal with the most *burdensome* and urgent healthcare needs of a given population? For instance, of what direct use are reproductive biotechnologies in a population whose main health concern lies in diminishing the numerous instances of malaria? How *feasible* is the actual application of these biotechnologies within specific cultures? Are we talking about an immediate application, or one that will occur in the long term? All technologies utilize new forms of *knowledge*. How realistic is the application of certain biotechnologies if there is a wide discrepancy in the knowledge base of the population? How well informed will a given population become with respect to certain biotechnologies? In her essay, Stella Gonzalez-Arnal addresses this in terms of the need to acknowledge a plurality of knowledge systems. What about *other benefits* besides health care that impact upon a population's health? Will biotechnologies improve the economy? Will they help to protect the environment? These are questions that must be addressed in order to assess the value and effectiveness of biotechnologies within specific cultural frameworks.[3]

What voices do we usually hear regarding biotechnology? When we review biotechnology's historic cast of characters, we see that biotechnology has been the product of scientists who, upon first glance, seem to represent voices from around the world. They include Robert Hooke, the English scientist who discovered the "cell"; Anton van Leeuwenhoek, the Dutch lens grinder who designed the microscope; the German biologists Matthias Schleiden, known for his cell theory in plants, and Theodore Schwann, who applied this theory to animals; the Englishman Charles Darwin and his adventures in the Galapagos Islands; the Austrian monk Gregor Mendel and his laws of heredity; the Swiss chemist Johann Miescher who discovered deoxyribonucleic acid (DNA), the building blocks of genetic material; the British scientist Fred Griffith who first noted the movement of genetic material to different strains of bacteria; the Canadian bacteriologist and physician Oswald Avery who discovered that DNA carried hereditary information; the American geneticist George Beadle and the American biochemist Edward Tatum, with their research on genes and enzymes; the American biochemist James Watson and the British biophysicist Francis Crick and their discovery of the double helix of DNA; and the Pakistani-American biochemist Har Gobind Khorana and the American biochemist Marshall Nirenberg who finally solved the genetic code.[4] Upon closer inspection, however, we see that this international collaboration has been more Western. Where are the voices from non-Western cultures? Listening to these voices is all the more critical since the influence of the above scientists has been profoundly global. We are all stakeholders in the work of biotechnology.

The voices we do hear today come from scientists, public policy experts, academicians, and, of course, biotech industry representatives who naturally portray biotechnologies in a favorable light. And there are certainly voices of dissent. Yet, where are the voices from other cultures, other traditions? The fact of the matter is that the voices we hear do not tell the whole story. Both trumping the benefits of biotechnologies and pointing out its deficiencies serve a purpose, but not when the perspectives from other cultures remain hidden.

This book rests upon at least two fundamental premises. The first is that it is entirely naïve on our parts to assume that the values we place upon certain biotechnologies are also held by all other cultures. We are learning this lesson the hard way in health-care ethics. Many Western cultures may cherish the principle of individual autonomy along with individual rights and decision making. Yet we are now facing the hard reality that this notion is not necessarily shared by most of the world. David Chan points this out clearly in his insightful discussion of the need for informed consent in Asian countries such as Singapore, with its strong tradition of Confucianism.

Must we therefore conclude that the rest of the world is wrong? Should we see to it that other cultures come to accept the values and principles we cherish? Do we have a moral obligation to bring justice, or at least our version of justice, to the rest of the world? Or is this an expression of moral hubris on our parts? Is this yet another example of what critics label Western imperialism?

As a critical first step in order to avoid this charge of imperialism, we need to critically examine the views of biotechnologies within other cultures. This entails recognizing how specific biotechnologies are applied in these cultures, as well as understanding the unique nature of the tensions that this creates within cultural contexts. Unless we make a genuine effort to examine biotechnologies within the framework of diverse cultures, our most ardent detractors will be justified in their criticism that we are in effect ethical imperialists, and biotechnology is our newest, long-range weapon.

Clearly, there is much at stake in all this. In order to more properly evaluate the numerous issues, we need to gain a clearer picture of both benefits and risks of biotechnological applications. At the same time, we need to situate these applications within their cultural contexts and assess whether or not there is sufficient sensitivity to these cultures' values and principles. Are we sufficiently respectful of the diversity within our global community and all that this respect for diversity entails? These are the leading questions that take us into the twenty-first century, and they act as a moral barometer on our biotechnological imperatives.

Given the interests of the numerous stakeholders in biotechnologies, here is the ultimate challenge: Can we cultivate a shared vision of the goals and applications of biotechnology? This need for a shared vision is the second key premise behind this book. The text offers an invaluable array of essays by experts in fields directly related to biotechnologies. The authors particularly address applications in agriculture and health care, the two most far-reaching areas impacted by biotechnology. In many respects, this is a primer on biotechnology for the uninitiated, a nicely compact introduction to an extremely complex subject. At the same time, the text is sophisticated enough to challenge those already familiar with biotechnology's terrain.

What is especially unique about this book is that it offers, in one concise volume, a variety of perspectives from different cultures. The idea for this book was stirred by an international gathering of scholars and experts at Salzburg, Austria, in October 2001. This Salzburg Seminar on "Biotechnology: Legal, Ethical, and Social Issues" assembled representatives from thirty different countries. It proved to be an immensely engaging experience that demonstrated all the more the need for cross-cultural dialogue regarding the numerous challenges posed by biotechnology. This book therefore challenges us in many directions. While the authors single out specific biotechnological applications along with accompanying concerns in their respective countries, they themselves represent diverse cultures. Yet these diverse representatives are all of one voice when it comes to the most important question: In our world community, what benefits can biotechnology produce for the world as a whole? What are the benefits of biotechnologies for the world's poor, starving, and those ravaged by disease? Can we distribute biotechnology's benefits in an equitable fashion? Does biotechnology offer more promise or peril?

NOTES

1. Global Forum for Health Research, *The 10/90 Report on Health Research 2000* (Geneva: Global Forum for Health Research, 2000).

2. Abdallah S. Daar et al., "Top Ten Biotechnologies for Improving Health in Developing Countries," *Nature Genetics* 32 (October 2002): 229.

3. Daar et al., 229.

4. Eric S. Grace, *Biotechnology Unzipped: Promises and Realities* (Washington, D.C.: Joseph Henry Press, 1997), 28–29.

I

INTERNATIONAL OVERVIEWS AND POLICIES

This opening section offers an overview of the major issues in biotechnology within a global and international framework. It also sets out a course for reflection, practical application, and action by laying out the fundamental issues in light of international policies and regulations.

As a start Don Chalmers nicely addresses critical questions concerning biotechnology's impact, appropriateness, and feasibility. He lays a key issue the increasing tension between the commercialization of biotechnology and its equitable application. Are the two incompatible? The biotechnology revolution is clearly determined by the "scientific power blocks" of the developed countries. They have their own agenda with respect to biotechnology's priorities, unfortunately, in many cases, an agenda that seems to conflict with the needs of poorer nations. The richer countries are the driving force behind the commercialization of biotechnologies, as evidenced by patenting applications. As Chalmers points out, this increases the gap between developed and developing countries.

Knowledge needs to be accompanied by responsibility for that knowledge. Yet Chalmers warns us that this has not been the case with biotechnology, particularly concerning human rights. The commercialization of biotechnology has in effect sustained an ever-widening gap between the rich and the poorer nations, and this necessitates serious public discussion and debate. Legislation, such as Australia's legislation regarding genetically modified organisms, needs to be judicious. There need to be international steps toward equitable benefit sharing. The 1992 Convention on Biological Diversity is a necessary first step. At the same time, multinationals must also be committed to the vision of shared benefits.

There is perhaps no more profound example of biotechnology's global repercussions than the International Human Genome Project (HGP). As Michael Morgan and Susan Wallace point out, its effects will be astounding, enabling us, for the first time in human history, to have more of a direct say in our destinies. They take this issue up in an eminently informative and insightful fashion, and give us an excellent introduction to this extremely complex topic. Not only do they provide an exceptionally clear survey of the HGP's history and political dimensions, but they also indicate how this international project raises questions about who and what we are, as well as the intricate ties between science and society.

The lessons from the project are indelible, particularly in terms of its numerous potential benefits in areas such as expression profiling, proteomics, and structural genomics. And they again remind us, as does Chalmers, that the true challenge in this biotechnological marvel lies in ensuring equitable benefits to all peoples. How can we bring about affordability? How can we apply research to address the most pressing health-care burdens in poorer countries? How can we prevent the abuse of this knowledge including the abuse of patenting? Acknowledging these ethical issues requires partnerships on all levels, particularly international. As they aptly put it, the benefits from studying the genome "belong to everyone."

As former chair of the United Nations Educational, Scientific, and Cultural Organization (UNESCO), Ryuichi Ida also gives us an overview of ethical and legal concerns raised by biotechnology within an international, global framework. At its most basic level, biotechnology represents a monumental victory in science. Yet, at the same time, its repercussions are so profound that it raises a host of ultimately serious questions such as redefining the meaning of human life and human value.

Ida addresses three areas of primary concern: human dignity, human rights, and economy. And he does so by examining specific fields within human genetic research, such as human reproductive cloning and research utilizing human embryonic stem (ES) cells. He also situates these concerns within international instruments such as UNESCO's 1997 Universal Declaration on the Human Genome and Human Rights. Furthermore, he singles out Japanese official statements like the discussion from the Bioethics Committee of Japan. Ida concludes by underscoring economic considerations including those pertaining to intellectual property rights. Ultimately, unless there is a concerted international effort to address these issues and to restore the notion that science must be at the service of society and culture, the victory of biotechnology will ring hollow.

1

Commercialization and Benefit Sharing of Biotechnology: Cross-Cultural Concerns?

Don Chalmers

THE BIOTECHNOLOGY REVOLUTION

Following the agrarian, industrial, and petroleum revolutions of prior centuries, the current biotechnology revolution will no doubt dominate this century. As the High Court of Australia's Justice Michael Kirby puts it, "the most important scientific breakthrough of [the twentieth] century may be seen, in time, to be neither nuclear fission, nor interplanetary flight, nor even informatics, but the fundamental and basal molecular biology which permits the human species to look into itself and find, at last, the basic building blocks of human and other life."[1] And while the former revolutions had profound effects on all nations, this biotechnology revolution remains concentrated within the scientific power blocks of North America, Europe, Japan, and other developed nations. Patent applications are a fair measure of this concentration. For example, in 1991, there were approximately four thousand patent applications worldwide for expressed sequenced tags. In 1996, there were some twenty-two thousand, mostly concentrated in the United States and Europe. By 2000, the U.S. Patent Office received around twenty-five thousand genomics applications.[2]

Developed nations are cultivating genetic and biotechnological horizons. These nations have taken steps to promote research and increase research funding and support for start-up companies to drive the biotechnology revolution. They have even introduced reforms to ensure that the taxation regimes for biotechnology companies are consistent with other Organisation for Economic Cooperation and Development (OECD) nations as a means of encouraging investment. It is hoped that advances in biotechnology will improve the health of these nations' populations as well as build their

economies to create valuable jobs. Australia, for example, is committed to a
biotech future and has declared in its national strategy that "consistent with
safeguarding human health and ensuring environment protection, that Australia capture the benefits of biotechnology for the Australian community,
industry, and environment."[3]

Unfortunately, the developing nations have little or no part to play in this
revolution. And unless benefits are shared and technologies are transferred,
the commercialization of biotechnology will indeed further deepen the gulf
between the developed and developing world. In his *Wealth of Nations*,
Adam Smith argued that mercantilism entailed a moral responsibility to bring
about benefits for all. Smith's eighteenth-century view must be revisited.
Herein lies the challenge: to implement policies and practices that ensure
that biotechnology's benefits are enjoyed by all, including the underprivi-
leged in the developing world.

BIOTECHNOLOGY IN PRACTICE

The biotechnology revolution has resulted in significant strides in the areas of
pharmaceuticals and agriculture. For instance, the Human Genome Project
(HGP) has already completed the mapping of the entire human genome. This
significant event marks the beginning of further extensive research into disease identification and treatment as well as new pharmaceuticals. The language of genetics has demonstrated a subtle journey to complexity over the
last ten years: from monogenetic to polygenetic to multifactorial disease analysis; from [looking at a] single gene to many genes to genes linked with environmental factors. With the gene as both "a discrete structure and a continuous variable,"[4] Dr. Craig Venter now claims that the idea that one gene
produces one protein or equals one disease is rapidly "flying out of the window."[5] Jeroboams of scientific and academic ink are now devoted to the many
offshoot disciplines of the HGP, particularly structural genomics (gene identification), proteomics (protein structure), functional genomics (genetic variation), transcriptomics (gene expression monitoring), targeted drug discovery
and pharmacogenomics for treatment of diseases with a genetic component,
and the enabling technologies of microarray genechips and bioinformatics.[6]

Due to biotechnology's plant and animal research, molecular biology has
accelerated the capacity for plant and animal manipulation on the farm. At
the same time, this research has raised specific concerns regarding genetically modified organisms (GMOs) in the agricultural and primary industries
sectors. The considerable effort expended crossing human genes into animals, particularly pigs, in order to produce new medicines and vaccines
evokes Steve Jones's claim that biotechnology is "interfering with nature."[7]
With serious efforts to develop compatible transplantable animal parts

through transgenics, this xenotransplantation[8] is being viewed as a means for replenishing the diminishing supply of human body parts for transplantation and substituting scarce supplies of whole blood and blood products. However, there are concerns that xenotransplantation may pose a grave risk to the community through the possibility of introducing animal viruses or new pathogens into human beings.

Since there are real risks of pandemic outbreaks,[9] most regulators and scientists embrace a "precautionary principle" requiring that progress in this area be slow and cautious in order to assess and avoid potential risks.[10] Yet there are profound differences when it comes to interpreting this principle.[11] One company in particular, Novartis, continues to invest heavily in this area of research and allegedly controls half of the worldwide transplantation drugs and xenotransplants.[12] And although the problems of hyper-rejection of organs in efforts of animal to human transplants have been insurmountable, enormous amounts, estimated at $1 billion annually, have been invested, particularly in relation to developing genetically modified pigs as a source of animal parts for compatible human transplantation.[13]

HUMAN RIGHTS

In 1994, the United Nations Economic and Social Council declared that "the rapid development of science and its technological applications . . . have not been accompanied by an appropriately urgent, profound and continuous consideration of their implications for human rights."[14] Similar sentiments have been expressed by other international voices: the 1993 papal encyclical *Veritatis Splendor*, the Council of Europe's protocol on cloning, and the World Health Organization.[15] And the United Nations Educational, Scientific, and Cultural Organization (UNESCO), in its Declaration on the Human Genome and Human Rights, included two articles about personal liberties:

> Art. 6   No one shall be subjected to discrimination based on genetic characteristics that is intended to infringe or has the effect of infringing human rights, fundamental freedoms and human dignity.
> Art.10   No research . . . concerning the human genome . . . should prevail over respect for the human rights, fundamental freedoms and human dignity of individuals or . . . groups of people.[16]

The 1990s witnessed an increased awareness of genetic knowledge and its challenge to human rights, such as potential stigmatization and discrimination through genetic "labeling."

Human rights issues were especially prominent in the debates over human cloning. The announcement of Dolly, the cloned sheep, unleashed a torrent of international concern about the capacity of science to convert *Homo sapiens* to *Homo proteus* through genetic intervention.[17] As UNESCO's Declaration on the Human Genome and Human Rights states in Article 11, "[p]ractices which are contrary to human dignity, such as reproductive cloning of human beings shall not be permitted."[18] Similarly, to its Convention on Human Rights and Biomedicine the Council of Europe added its Protocol on Prohibition on Cloning of Human Beings, declaring that "Any intervention seeking to create a human being genetically identical to another human being, whether living or dead, is prohibited."[19] Many European countries formed their own legislation prohibiting any cloning intended to produce genetically identical individuals. Much of this legislation was based on perceived affronts to "human dignity," a term that bioethicist John Harris criticized as being "comprehensively vague."[20] Meanwhile, in the United States, the National Bioethics Advisory Commission Report, "Cloning Human Beings," recommended legislation combined with a moratorium and cautioned that "any regulatory or legislative actions undertaken to effect the foregoing prohibition on creating a child by somatic cell nuclear transfer should be carefully written so as not to interfere with other important areas of scientific research."[21] These other areas included stem cell technology, sometimes referred to as "therapeutic" cloning.[22] Similarly, the United Kingdom issued reports that supported licenses by the Human Fertilization and Embryology Authority to allow research on human parts to investigate the potential benefits of the techniques to mitochondrial diseases and damaged tissues and organs.[23] In all this, it is important to bear in mind that these concerns about genetic discrimination and human cloning are firmly rooted in the individualistic notion of Western human rights. As such, these concerns do not embrace the more systemic problem of exclusion of the developing nations from an equitable share in the benefits of the biotechnology revolution.

PUBLIC OPINION AND CONSULTATION

By no means has public enthusiasm for the biotechnology revolution been wholesale. Developed countries exude confidence as to medical applications of biotechnology, but show reservations about agricultural uses. More importantly, in the developed world, there has been virtually no discussion about the ethical responsibilities of sharing benefits with developing nations. Attitudes in the developed world can be gauged from the regular polling of public opinion within the European Union (EU). The EU polled sixteen thousand people in its survey, "The Europeans and Biotechnology," and key findings were:

- More than 80 percent of Europeans feel poorly informed about biotechnology, but most are willing to learn.

- For nearly half, "biotechnology" prompts ideas of cloning animals and humans, thus eliciting strong negative emotions, while research, health, and environment issues are viewed positively.
- A majority think that technologies such as solar energy, information technology, telecommunications, and the Internet "will improve life in the next twenty years." Only 41 percent (down from 46 percent in 1996) think the same for biotechnology. Only nuclear power (26 percent) attracts less confidence.
- There is a consensus that it is morally acceptable to use genetic tests to detect inherited diseases, to develop genetically modified bacteria to clean pollution, and to introduce human genes in bacteria to produce medicine or vaccines. There was more limited acceptance of cloning human cells or tissue to help a patient and to transfer plant genes to other plants in order to obtain resistance to insects. There was far less support for the use of biotechnology in food production to improve taste or nutritional content and in the cloning of animals, even for medical applications.
- There is considerable suspicion of public authorities and technical experts. In response to a question about which sources of biotechnology information are trustworthy, consumer organizations (55 percent), the medical profession (53 percent), and environmental protection organizations (45 percent) fared best. Universities (26 percent), animal protection organizations (25 percent), and the media (20 percent) had modest levels of support. There is a high degree of skepticism about international institutions (17 percent), national public authorities (15 percent), farmers' associations (15 percent), and religious organizations (9 percent). Levels of trust for all sources of biotechnology information decreased significantly since the last survey in 1996.
- Only 45 percent feel that their governments regulate biotechnology well enough.[24]

The biotechnology revolution has been accompanied by unprecedented public consultation through national commissions of enquiry, particularly in the area of agricultural biotechnology. In France, the National Consultative Ethics Committee for Health and Life Sciences has been operating since 1983 and gives opinions on problems raised in the fields of biology, medicine, and health. Australia has set up the Australian Health Ethics Committee, which is required by statute to conduct a two-stage public consultation on any matter in human genetics referred to it for consideration. Similarly, the United States established its National Bioethics Advisory Commission to make recommendations on human experimentation and human genetics. The United Kingdom instituted the Human Genetics Advisory Commission, which, inter alia, promotes public debate and discussion on genetics issues.

The Danish Council of Ethics has produced much distinguished work. It should also be noted that Article 28 of the Convention on Human Rights and Biomedicine, promulgated by the Council of Europe in November 1996, requires parties of the convention to "see to it that the fundamental questions raised by the developments of biotechnology and medicine are the subject of appropriate public discussion in the light, in particular, of relevant medical, social, economic, ethical, and legal implications, and that their possible application is made the subject of appropriate consultation." However, and this is critical, these discussions have essentially been conducted in and for the developed world, even though there have been a few notable exceptions such as the work of the Nuffield Council of Bioethics in its report on "The Ethics of Research Related to Health Care in Developing Countries."

REGULATION OF BIOTECHNOLOGY

Many developed countries have introduced legislation dealing specifically with GMOs. Let's take the case of Australia, which originally had in place a voluntary system of regulation of GMOs since 1975, under the guidance of the Genetic Manipulation Advisory Committee (GMAC) and its predecessors. There appears to have been a high level of compliance with GMAC recommendations. However, GMAC operated within an administrative system, with no legally enforceable auditing or monitoring of compliance and no legal basis for the imposition of penalties or other action in the event of noncompliance. Industry had concerns about the lack of rules and standards, creating uncertainty in the market. In 1998, the deficiencies in Australia's voluntary system of regulation of GMOs were recognized. Federal, state, and territory governments then collaboratively crafted legislation after public consultation.

Australia's Gene Technology Act 2000 came into effect on June 21, 2001, and applies to all dealings with GMOs including experimentation, production, breeding, and importation of a GMO, or using a GMO in the manufacture of another thing. The act also instituted the Office of the Gene Technology Regulator, which has the primary role in regulating dealings with GMOs. The higher the risk factors involved in a particular dealing, the greater the level of regulation. Thus while some dealings are exempt from regulation, others (notifiable low-risk dealings) must be reported to the Gene Technology Regulator, conducted in an accredited facility by an accredited organization, and cannot be released into the environment. Any dealings of a higher risk, including those involving intentional release into the environment, can only be conducted under a license granted by the regulator. The applicants themselves must present risk assessment plans. Moreover, the regulator is excluded from any consideration of economic issues in assessing risk and deciding whether to grant a license. Licensed dealings that have proven to have no environmental risk are then placed on the GMO Register. Once listed on the GMO Register, dealings can thereafter be undertaken by any person or organization *without* license.

There are also opportunities for the public to comment on both the applicant's risk assessment as well as the risk management plan. Further public involvement may come about through three advisory committees that counsel the regulator on science, ethics, and the community. It is important to bear in mind that the regulator has neither extraterritorial powers nor any role in relation to developing countries.

COMMERCIALIZATION

By the close of the last century, the "great race" between the international collaborators in the Human Genome Project and the private U.S. company, Celera Genomics, to sequence the human genome clearly symbolized the exponential growth in biotechnology's commercialization.[25] The estimated $3 billion of public money spent on the Human Genome Project was augmented by private billions on commercial research and product development.[26] By late 1998, there were some two hundred thousand people working in more than one thousand biotechnology companies, not including academic and government scientists. By 2000, there were 1,273 biotech companies in the United States and 1,351 in Europe. The market is estimated to expand to $6 billion by 2010.[27] And there continues to be major industry buy ups of smaller companies working on drug targets.[28]

Patenting in the research and development area posed special problems. As Jasanoff has noted under patent law, benefits are "bestowed on the first to discover a new prize or invent a new product. Prizes go to the first to publish. . . . Science . . . is a winner-take-all game, with no glory or comfort for the also-ran."[29] For example, the patenting of genes is a hotly discussed and disputed issue.[30] Although patents carry the usual guarantee of protection for investment in new technologies, they may not be appropriate in the area of human genetics. Here, there is a clear divide between two camps. On the one hand, many scientists believe that gene identification is simply the discovery of something that already exists and is therefore not patentable and should remain public domain. On the other hand, many claim that this still amounts to an invention that is patentable just like any other process. The patentors are clearly prevailing. In 2000 the gene patent rush included around 126,000 applications for whole or partial human gene sequences. Many of these human gene sequence applications have been successful *where their function is known.* Furthermore, although this dispute is concentrated in the developed world, it has an enormous effect on the developing nations.

Note the rapidly rising rate of patenting application. In October 2000, the patent rush included 9,364 patent applications filed worldwide for inventions related to the human body and some 126,672 applications for whole or partial human gene sequences. By the end of 2000, gene sequence applications increased by another 34,500 applications. Incyte Genomics has filed applications covering portions of more than 50,000 gene sequences including 7,000 full-length genes and has already been granted 500 gene-related patents.[31] Facing a backlog of applications, the U.S. Patent and Trademark Office has declared that they will be applying a more stringent test of "utility," hoping to prevent the bestowal of patents to inventors who have little idea how the so-called "invention" functions. In this way, the U.S. office is coming closer in line with that of the EU.

It has been estimated that the gap between what a person produces in the rich and poor nations is now as wide as 390:1.[32] Along these lines, greater commercialization and patenting are further alienating the developing nations. The United States and Europe control the biotechnology industry and dominate its markets in technology, therapeutics, new generation vaccines, gene therapy, diagnostics, and new pharmaceuticals (molecular "pharming"). Despite this control by the developed nations, the World Trade Organization Agreement on Trade-Related Aspects of Intellectual Property Rights (TRIPS) arguably seems to favor developing countries in achieving greater self-reliance in biotechnology. Jayashree Watal has acknowledged that intellectual property rights for biotechnological inventions "do pose complex problems relating to access to technologies, unfair exploitation of genetic resources and equitable sharing of the financial benefits." At the same time, she asserts that "liberal use should be made of the flexibility available under TRIPS to grant compulsory licences" and suggests that the best way forward for developing countries should be through "collaboration and confrontation."[33]

BENEFIT SHARING

The gap between rich and poor has never been so wide, nor widening at such a rapid pace. Within the developed world's biotechnology horizon, the challenge is to ensure that this horizon includes the developing nations through an equitable distribution of the benefits of this revolution. The extraordinary promise of genetics "carries with it extraordinary responsibilities. It is incumbent on both scientists and public servants to ensure that science serves humanity always, and never the other way round."[34] Yet the high costs involved in the developments and applications of biotechnology may lead to exclusion of the poor and needy from access to the products. Lee Silver paints a stark contrast between the "gen rich" who are able to afford the expense of new technologies and the "naturals" who will be forced to use cheaper versions of the technology or else be excluded from the benefits altogether.[35] The United Nations Economic and Social Council concluded that "[t]here is a general recognition of the need for international cooperation in order to ensure that mankind as a whole benefits from the life sciences and to prevent them from being used for any purpose other than the good of mankind."[36]

The Human Genome Organization (HUGO) made a significant effort to consider this issue of benefit sharing. HUGO's Ethics Committee addressed issues specifically related to genetic research in developing countries and the transfer of benefits during such research, leading to the publication of its Statement on Benefit Sharing. This document differentiated a benefit and a profit in the monetary sense. The determination of a benefit depends on needs, values, priorities, and cultural expectations. The document recommended that in the case of profit making, the profit-making entity should dedicate a percentage of its annual net profit, for example 1–3 percent, to healthcare infrastructure or to humanitarian efforts for the group.[37]

International law has also established the principle of benefit sharing in the area of biodiversity and genetic resources in food and agriculture. When world leaders met in 1992 at Rio de Janeiro for the Earth Summit, they formulated a pact known as the Convention on Biological Diversity. The convention proposed procedures for fair and equitable distribution of benefits and for the appropriate transfer of new biotechnological technologies to developing countries. In particular, the convention states that "Each Contracting Party shall take legislative, administrative or policy measures, as appropriate, . . . with the aim of sharing in a fair and equitable way the results of research and development and the benefits arising from the commercial and other utilization of genetic resources with the Contracting Party providing such resources" (Article 15, "Access to Genetic Resources"). In addition, the convention also extends the benefit sharing ideals to technology transfer. In particular, Article 16, "Access to and Transfer of Technology," provides that

1. Each Contracting Party . . . undertakes . . . to provide and/or facilitate access for and transfer to other Contracting Parties of technologies that are relevant to the conservation and sustainable use of biological diversity or make use of genetic resources and do not cause significant damage to the environment.

2. Access to and transfer of technology referred to in paragraph 1 above to developing countries shall be provided and/or facilitated under fair and most favourable terms. . . .

3. Each Contracting Party shall take legislative, administrative or policy measures, as appropriate, with the aim that Contracting Parties, in particular those that are developing countries, which provide genetic resources are provided access to and transfer of technology which makes use of those resources, on mutually agreed terms, including technology protected by patents and other intellectual property rights, where necessary . . .

Article 19 of the convention deals with "Handling of Biotechnology and Distribution of its Benefits" and declares that "Each Contracting Party shall take all practicable measures to promote and advance priority access on a fair and equitable basis by Contracting Parties, especially developing countries, to the results and benefits arising from biotechnologies based upon genetic resources."

Finally, the Cartagena Protocol (2000) to this convention included obligations with respect to capacity building through cooperation "in the development and/or strengthening of human resources and institutional capacities in biosafety, including biotechnology to the extent that it is required for biosafety, for the purpose of the effective implementation of this Protocol, in developing countries Parties, in particular the least developed and small island developing States among them, and in Parties with economies in transition . . ."[38]

CONCLUSION

The biotechnology revolution has the potential to further widen the gap between the developed and developing world. As much as biotechnology is a transnational revolution, so too biotechnology should support partnerships between the developed and developing nations to secure an equitable distribution of the benefits of this revolution. The articles of the Convention on Biological Diversity are a sound foundation for further commitment of nations to benefit sharing. However, in reality it will be multinationals like Merck, Microsoft, and Monsanto that have to be committed to this vision. And regulation needs to be introduced within a framework that is internationally consistent. The regulations in one country clearly impact upon those of another. Inevitably, the research agenda will be set by the developed world. Unsurprisingly, the developed world is more interested in the genetics of cancer and cardiovascular diseases than in cures for the Third-World scourge of malaria. Nevertheless, recent pronouncements such the Convention on Biological Diversity need to be embraced by those drawing on the rich promises of the biological revolution. Otherwise, the revolution will become an enemy of its own promise.

NOTES

1. M. Kirby, "Legal Problems: Human Genome Project," *Australian Law Journal* 67, no. 894 (1993): 903.
2. J. Enriquez, *As the Future Catches You* (New York: Crown Business, 2000), 183.
3. Commonwealth of Australia, *Australian Biotechnology: A National Strategy*, 2000.

4. E. Fischer, "What's in a Gene?" in *Genethics* (Basel, Switzerland: Ciba-Chiron, 1991).

5. C. Venter, "Gene Watch," *Financial Times* 14 (2001): 5.

6. D. Nicol and J. Neilsen, "The Australian Medical Biotechnology Industry and Access to Intellectual Property: Issues for Patent Law Development," *Sydney Law Review* 23 (2001): 347.

7. S. Jones, *The Gene Genie* (U.K.: Sherwell, 1994).

8. M. Fox and J. McHale, "Xenotransplantation: The Ethical and Legal Ramifications," *Medical Law Review* 6 (1998): 42. See also A. Daar, "Ethics of Xenotransplantation: Animal Issues, Consent, and Likely Transformation of Transplant Ethics," *World Journal of Surgery* 21 (1997): 9.

9. D. Butler et al., "Briefing: Xenotransplantation," *Nature* 391 (1998): 320–328.

10. C. Patience et al., "Infection of Human Cells by an Endogenous Retrovirus of Pigs," *Nature Medicine* 3 (1997): 282.

11. J. Morris, ed., *Rethinking Risk and the Precautionary Principle* (Oxford: Butterworth-Heinemann, 2000).

12. *Genetic Engineering News* 18, no. 1 (1998): 31, 39.

13. *Nature* 391 (1998): 320; estimate is by Société Générale Strausse Turnbull.

14. U.N. Commission on Human Rights, 51st Session, *Human Rights and Scientific and Technological Developments*, Section 3, Conclusions and Recommendations (1994), paragraphs 152, 154.

15. WHO Resolution of the 50th World Health Assembly, Geneva, affirming that the use of cloning is ethically unacceptable and contrary to human integrity and morality (1997).

16. UNESCO:1997.

17. L. Silver, *Re-making Eden* (U.K.: Weidenfeld & Nicolson, 1998).

18. UNESCO:1997.

19. EC, *The Protocol on the Prohibition of Cloning Beings*, adopted by the Parliamentary Assembly of the Council of Europe, September 22, 1997, to the *Convention for the Protection of Human Rights and Dignity* with regard to the application of Biology and Medicine, November 1996.

20. J. Harris, "Goodbye Dolly? The Ethics of Human Cloning," *Journal of Medical Ethics* 23 (1997): 353.

21. NBAC2, National Bioethics Advisory Commission Report, *Cloning Human Beings*, (Rockville, Maryland, 1997), Recommendation III.

22. J. Robertson, "Human Cloning and the Challenge of Regulation," *New England Journal of Medicine* 339 (1998): 119; George Annas, "Why We Should Ban Human Cloning," *New England Journal of Medicine* 339 (1998): 122.

23. HFEA, a joint committee of the Human Genetics Advisory Commission and the Human Fertilization and Embryology Authority, *Cloning Issues in Reproduction Science and Medicine*, 1998.

24. International Research Associates (INRA) Europe–European Consumer Safety Organization (ECOSA), *Eurobarometer 52.1 (2002): The Europeans and Biotechnology*, March 15, 2000, europa.eu.int/comm/research/quality-of-life/eurobarometer .html.

25. M. Otlowski, D. Nicol, and D. Chalmers, "Consent, Commercialization and Benefit-Sharing," *Journal of Law and Medicine* 9 (2001): 80.

26. GenBank-National Institutes of Health, www.ncbi.nlm.nih.gov/GenBank/index .hmtl.

27. *Nature*, 1998.

28. Nicol and Neilsen.

29. S. Jasanoff, "Innovation and Integrity in Bio-medical Research," *Academic Medicine* 68 (1993): 95.

30. R. Eisenberg, "Patenting the Human Genome," *Emory Law Journal* 39 (1990): 721; D. Nicol, "Should Human Genes be Patentable Inventions under Australian Law?" *Journal of Law and Medicine* 3 (1996): 231.

31. Nicol and Neilsen.

32. Enriquez, 24.

33. J. Watal, "Intellectual Property and Biotechnology: Trade Interests of Developing Countries," *International Journal of Biotechnology* 2, no. 44 (2000): 55.

34. W. Clinton, "Catalyzing Scientific Progress," *Science* 279 (1990): 1111; transcript of speech, www.aaas.org/meetings/scope/clinton.htm.

35. Silver.

36. U.N. Commission on Human Rights.

37. HUGO, Human Genome Organization Ethics Committee, *Statement on Benefit-Sharing*, 2000, www.hugo-international.org/hugo/benefit.html.

38. *Cartagena Protocol,* Article 22.1.

2

The International Human Genome Project: An Overview

Michael J. Morgan and Susan E. Wallace

The human genome is the stuff that carries the information from one generation to another that makes us human and similar in many respects to our fathers, mothers, and our forebears. For almost all creatures on this earth that information is encoded in a beautiful molecule called deoxyribonucleic acid (DNA), a double helix whose structure was determined fifty years ago by Francis Crick, James Watson, and Maurice Wilkins. And it is the project to decode the DNA of humans and other organisms that has changed the world by giving us new scientific information, new research avenues to explore, and new insights into ourselves as human beings. It has also heralded a new appreciation that science and society are inextricably linked, and that the public needs to be engaged at all levels if scientists are to continue to be able to carry out their trade.

THE SCIENCE

The structure of DNA immediately suggested to Crick and Watson how genetic information might be transmitted from one generation to the next. The molecule consists of two interwoven helices consisting of a backbone (deoxyribosephosphate) bearing four nucleotides or bases. The bases have an interesting and essential chemical bonding property: adenine (A) can only pair with thymine (T), and cytosine (C) can only pair with guanine (G). This leads to the unique property of the individual helices (strands) being able to pair exactly with their complementary partner. Thus, if the sequence on one strand of the double helix is GATTAC, then the complementary strand's sequence is CTAATG, and this is true whatever the length of the molecule. To generate a copy (replica) of the double helix, the helix could unwind and each strand could serve as a template to reproduce an exact replica of the parent double helix. This simple semiconservative replication explains how genetic information can be accurately reproduced. The genome of humans actually consists of forty-six chromosomes, twenty-three coming from each parent, carried in the nuclei of essentially all cells. Each of these chromosomes consists of a single strand of DNA, a very thin, but long, molecule. If stretched out, the DNA of a single chromosome set in a single human cell would measure two meters in length. Stitch all of the DNA together from the billions of cells in a single human being and it would stretch from the earth to the sun and back three times!

The two-meter long virtual DNA molecule contains the entire human genome, made up of approximately three billion bases (nucleotides). Furthermore, the human genome can be read like a book. Just like letters in a certain order can be read as words, the order of bases (e.g., ATGC reads as codes for genes. The genetic code is then "translated," through a complex series of reactions, to produce gene products, proteins, that make up much of the human body. Muscle is made up of proteins, as is the pigment that carries oxygen in the blood. All reactions in the body are effected through special proteins called enzymes. Thus the principal role of the genes is to provide the instructions to make proteins. When the message goes wrong, a disease might be the result. Faulty proteins have been found to be responsible for hemophilia, sickle cell disease, cystic fibrosis, breast cancer, and many other diseases. A faulty protein may be inherited from generation to generation or might be a random occurrence.

The objective of the Human Genome Project (HGP) was to determine the exact position of each base in the human genome, finishing with the complete sequence of bases from one end of each chromosome to the other. A useful analogy is to think of mapping the Earth. The world can be seen as a genome, with each chromosome a country, each chromosome fragment a city, every gene a street, and the bases houses on that street. In other words, the project intended to determine the position of every house on the planet, who lives on which street, and what each person does. A working draft of the genome was completed in 2001 and the completion of the project was announced in April 2003.

HISTORY AND POLITICS

In 1987, James Watson (who with Francis Crick won the Nobel Prize for discovering the double helical structure of DNA) appeared before the U.S. Congress to request a budget of $3 billion to sequence the human genome. The Wellcome Trust, based in the United Kingdom, became involved in human sequencing in 1992 when it awarded research money to Dr. John Sulston to work at the newly created Sanger Center in Cambridgeshire. In 1996, when it became clear that no single country could do the project alone, the Wellcome Trust organized a meeting, in Bermuda, of all interested parties to discuss the organization of the project, how the work could be split up between the laboratories around the world, and how the data might be made available to the wider scientific community. The countries and organizations involved at the time included the Wellcome Trust, the British Medical Research Council, the U.S. National Human Genome Research Institute, the U.S. Department of Energy, and representatives of the French, German, and Japanese Genome Projects.[1] At the meeting participants agreed that it would benefit laboratories worldwide if raw (unedited) sequence data, two kilobases in length, were released every twenty-four hours into a publicly accessible database.[2] More importantly, no patents would be sought on those submissions. This was revolutionary stuff, and it was welcomed by scientists as well as funding agencies. In this way, the laboratories involved in the HGP would receive the resources they needed to complete the job. In addition, those scientists not directly involved in the HGP but conducting research using genome sequence data would have unrestricted access to the raw material they needed to conduct their own research. This policy of releasing data with no prior claims made on it enabled research to thrive throughout the worldwide scientific community.

In 1998 a private U.S. company, Perkin-Elmer Corporation (now Applera Corporation), announced plans to start a sequencing company, Celera Genomics, that would sequence the human genome in less time (by 2001 instead of 2005) and for less money ($150 million as opposed to $3 billion) than the public project. Dr. Craig Venter, then Celera president, at the annual meeting of genome researchers at Cold Spring Harbor Laboratory (CSHL), New York, proposed that the public project throw in the towel and tackle the mouse genome instead of the human and let his company complete the task! In return for completing the project more quickly and more cheaply, Celera would compile the information into a database, access to which would be available only via subscription. The company also intended to patent some hundred or more genes.

The thought of the human genome held for sale in a private database was against the principles of the public project. The fear that the human genome would be withheld from those who needed it prompted Michael Morgan, with John Sulston, to fly to CSHL to announce at that same meeting that the Wellcome Trust had recently decided to double the budget of the Sanger Center in order to enable it to continue its work on the HGP. Additionally, the Wellcome Trust would consider funding the Sanger Center to complete sequencing of the entire human genome if other public funding sources dropped out. This was largely motivated by the Wellcome Trust's conviction that the most important aspect of the HGP is that the data be in the public domain rather than in private hands. The other funding agencies agreed and the project continued. However, in response to the new competition, the organizers of the public project decided to produce a rough "working draft" of the human genome by 2001. By this time in 1998, scientists had begun to see the benefits of the project and were eager for more sequence data. In order to better complete the task, a small number of laboratories, the "G5 group," took over the bulk of the large-scale production of human sequence. The G5 group included the U.S. laboratories at Washington University, Baylor University, and the Whitehead Institute, with the Joint Genome Institute, supported by the U.S. Department of Energy, and the Sanger Center in the United Kingdom.

During 1998 the sequence of the first animal genome, that of the nematode worm, *C. elegans*, was published. This was by far the largest genome to have been sequenced, but at one hundred megabases was still one-thirtieth the size of the human genome. Attempts were made in 1999 to broker a collaborative agreement between the public project and Celera, but it was not successful, largely because of the different approaches to data release and Celera's obligations to its shareholders to turn a profit. In 2000 the complete sequence of the fruit fly, *drosophila*, was published, as was that of the first entire human chromosome (chromosome 22).

Another important event was the declaration by President Bill Clinton and Prime Minister Tony Blair on the need to make human genetic information freely available. This marked the first time that the data release policies of the HGP were endorsed at a government level. Unfortunately, the media misinterpreted the statement to mean that there should be no patents in human genetics. This was incorrect. In their declaration, Clinton and Blair pointed out that intellectual property was important in order for discoveries to be developed and utilized, but that the raw data from which discoveries could be made should remain unpatented. Unfortunately, the statement had a detrimental effect on the biotechnology sector of the U.S. stock market, which suffered its largest ever one-day fall.

A key media event of 2000 was the joint announcement in June, telecast from both the White House and 10 Downing Street, on the completion of rough drafts of the human genome by both the public project and Celera. Early in 2001, the two groups had discussed publishing the results of their work in the same issue of one of the major scientific journals. However, Celera demanded that the journal publish their paper without having any sequence data deposited in a public database for open inspection. In an unusual move, the journal *Science* agreed to allow this, even though it is normally its policy to require that data supporting the conclusions of a scientific article are made available for public scrutiny. The scientists of the public project felt this was inappropriate and published their papers in *Nature*.

In April 2003 the public Human Genome Project partners declared the project finished as part of the celebrations to mark the fiftieth anniversary of the discovery of the structure of DNA. At the same time the heads of government of the six participating nations issued the following joint proclamation:

> We, the Heads of Government of the United States of America, the United Kingdom, Japan, France, Germany and China, are proud to announce that scientists from our six countries have completed the essential sequence of three billion base pairs of DNA of the human genome, the molecular instruction book of human life.
>
> Remarkable advances in genetic science and technology have been made in the five decades since the landmark discovery of the double-helix structure of DNA in April 1953. Now, in the very month and year of the 50th anniversary of that important discovery by Watson and Crick, the International Human Genome Sequencing Consortium has completed decoding all the chapters of the instruction book of human life. This information is now freely available to the world without constraints via public databases on the World Wide Web.
>
> This genetic sequence provides us with the fundamental platform for understanding ourselves, from which revolutionary progress will be made in biomedical sciences and in the health and welfare of humankind. Thus, we take today an important step toward establishing a healthier future for all the peoples of the globe, for whom the human genome serves as a common inheritance.
>
> We congratulate all the people who participated in this project on their creativity and dedication. Their outstanding work will be noted in the history of science and technology, and as well in the history of humankind, as a landmark achievement.
>
> We encourage the world to celebrate the scientific achievement of completing the Human Genome Project, and we exhort the scientific and medical communities to rededicate themselves to the utilization of these new discoveries to reduce human suffering.
>
> His Excellency Jacques Chirac,
>   President of the French Republic.
> The Honorable George Bush,
>   President of the United States of America.
> The Right Honorable Tony Blair, M.P.,
>   Prime Minister of the United Kingdom.
> His Excellency Gerhard Schroeder,
>   Chancellor of the Federal Republic of Germany.
> His Excellency Junichiro Koizumi,
>   Prime Minister of Japan.
> His Excellency Wen Jiabao,
>   Premier of the State Council of the People's Republic of China.[3]

WHAT HAVE WE LEARNED?

What has this project been all about? Dr. Francis Collins, one of the directors of the U.S. funding agencies supporting the project, has referred to the data as "a revolutionary textbook of medicine." By this he was anticipating that in the future the practice of medicine will become more and more rational based on a real understanding of the disease process. One example of this, which may impact patients soon, is the advent of "personalized" medicine. Through the Human Genome Project we have learned that at the level of DNA sequence we are all remarkably alike. All humans are 99.9 percent similar in our DNA, leaving 0.1 percent to differentiate us. This however still accounts for three million nucleotide differences. These differences, or single nucleotide polymorphisms (SNPs), may enable us to differentiate populations on the basis of the pattern of SNPs, which are associated with susceptibility or resistance to disease. These variable patterns of SNPs, also known as haplotypes, seem to be passed on from one generation to another. Efforts are under way to determine a common haplotype map, which will help diagnose the susceptibility of individuals to disease and determine which are the most appropriate drugs to treat those diseases. In order to further this research, the Wellcome Trust, in collaboration with members of the pharmaceutical industry and other companies, set up a consortium to fund a research project to identify human SNPs, with the resulting database in the public domain. If we can fully understand SNPs, one day it is likely that instead of going to your doctor to be prescribed a drug for hypertension that might or might not help you, a DNA test will tell your doctor which form of hypertension you have and you will be prescribed the drug for that type. Much morbidity in modern developed countries is due to side effects from drugs. If that burden could be relieved many more people could be helped.

THE RESEARCH AHEAD

The sequencing may be completed but that does not mean that there are not many more exciting projects ahead. The potential benefits and future research that may derive from the HGP are many, including gene therapy, new therapeutic targets and intervention strategies for medicine, the potential for identifying those at risk (predictive medicine), and a clarification of what it means to be human in relation to the other species on earth. The human genome sequence is a new language and we need to learn its grammar, syntax, and punctuation. In order to achieve a better understanding of our genome, science is now moving into functional genomics.

EXPRESSION PROFILING

It has been known for many years that the expression of genes varies from tissue to tissue and in normal versus diseased tissue. A single fertilized human egg has all the information it needs to turn us into an individual. During its development, certain genes are turned off and on, activating or deactivating certain proteins, enabling them to "make" skin tissue or brain tissue. Different genes are expressed in different ways for different purposes. Some genes are expressed to create kidney cells while different genes are expressed at perhaps a different time to produce liver cells. Similarly a liver cell that is normal has different genes expressed in it than one that is cancerous. If these differences can be identified, they can be targeted and scientists can learn what it is that underlies the disease process and, with skill and determination, design appropriate drugs. We are now able, through the development of gene "chips," to determine which of many genes are expressed in any tissue and in any metabolic state. Eventually we will be able to build a database showing which genes are expressed in any tissue at any time of development or pathological state.

PROTEOMICS

As discussed, genes are translated into proteins. One of the big surprises arising from the Human Genome Project is the revelation that we only have enough genes to code for around thirty thousand proteins rather than the one hundred thousand that had been predicted. Some people were quite upset to find out they were not much more complicated than a worm, an animal with approximately nineteen thousand genes! However it seems that mammalian genomes are crafty and can use the same stretch of DNA to code for several different protein products. Through this shuffling or alternative splicing, our bodies probably make more than one hundred thousand proteins from about thirty thousand genes. Proteins then get modified by the addition of various small molecules such as phosphate. Proteomics is the process of determining all the proteins in all the different cells and tissues, all of their variations and the ways they change in disease and health in time and space. This is an immensely challenging task that will take many years to comprehend.

STRUCTURAL GENOMICS

A protein is a very large organic molecule with a three-dimensional structure that determines its function. Scientists are using X-ray diffraction techniques and high-throughput automation to try to determine the structure of all human proteins. By comparing the structure of newly discovered proteins with a protein of known function, it is hoped that the function of the new protein can be understood. Many of these proteins will be new and important therapeutic targets and are of considerable interest to pharmaceutical companies who may then design new drugs. An international project is now under way to coordinate worldwide efforts to determine the structure of all proteins important to human and animal health.

DATABASES

The HGP has produced a large amount of information, and further studies will continue to add to our knowledge. The amount of DNA sequence, protein structure coordinates, and other biological data being deposited into databases by scientists is enormous and increasing exponentially. The computer industry is now driven by bioinformatics. However, researchers in academia and industry expect and are entitled to freely access the data to ensure the research is translated into new medicines or new ways of thinking. Therefore this endeavor must be in the public domain and internationally available. It cannot be restricted or owned by companies.

LIMITATIONS TO GENOMICS

Naturally there are limitations to genetic research. The timescales for achieving results are unclear and potential benefits may be seen by some as deterrents. There is a danger of overhyping the possible positive outcomes of the science. Other public health measures must be taken alongside genomics. Genomics also cannot, of itself, solve the problems of delivery of drugs to the poor. In addition, the public must be taught more about the science and its components and ramifications so that they can see it as it is—as research, not the Holy Grail or a potential monster.

THE FUTURE

The completion of the HGP in April 2003 is but a first step. There are huge challenges ahead to deliver the benefits of this science to the whole world. For example, how can genomics be used to treat disease in developing countries? The developed world surely has a responsibility to build a research capacity in developing countries so that they can address their specific problems. The cost of technology is high but there are many ways of adapting expensive technology to more inexpensive, though robust, alternatives. Regional and international partnerships are needed to increase access to this science. Although it is potentially contentious, genetic epidemiology (linking patient data with DNA samples) is going to be a major requirement in order to reap the full benefits of this science. For this to happen, researchers need access to populations around the world. Agencies in the United Kingdom, for example, are in the process of setting up a project to recruit five hundred thousand volunteers. The project will look at their DNA and their health outcomes across many years. However, the project is still under ethical review to ensure everyone is comfortable with how it will be done, what will be done with the data, and who will be able to access it. Once approved and begun, it is expected that the data will help in the development of medical products and interventions over the next decades.

The HGP has raised a number of ethical, legal, and social issues. This was recognized at the outset of the funding of the U.S. portion of the project. Dr. James Watson, then leading the U.S. project, determined to devote 3–5 percent of the overall HGP budget to study these issues. A number of ethical concerns will be addressed in this volume: Who owns DNA? How will we ensure that the benefits of the science are delivered to all the world's population? How will we prevent the science from being perverted for the benefit of the few through an application of a new eugenics? The science is moving so quickly that people fear being overwhelmed by it. There are also issues surrounding the patenting of genetic information. For instance, can the pressure to patent possibly corrupt researchers?

These are weighty issues that deserve the total engagement of society. This is just the beginning of a fundamental revolution in biology that is providing us with a profound knowledge and understanding that will, for the first time in our history, enable us to determine our own destiny. International involvement is necessary, as the science should not be controlled by a small group of nations or companies. The genome and all the benefits derived from studying it must belong to everyone.

NOTES

1. The Chinese joined the HGP at a later date.
2. A kilobase is a sequence consisting of one thousand bases.
3. U.K. Department of Trade and Industry, "Heads of Government Congratulate Scientists on Completion of Human Genome Project," April 12, 2003, at www.gnn.gov.uk/gnn/national.nsf (accessed September 5, 2003).

3

Ethical and Legal Aspects of Biotechnology

Ryuichi Ida

Biotechnology, the progress of science and technology relating to all living beings, particularly to human beings, was the twentieth century's most remarkable phenomenon. The new millennium is precisely an era that applies this promising new field of life science and biotechnology. In short, biotechnology in its broader sense represents a victory of human intelligence. However, at the same time, biotechnology poses grave challenges, particularly in relation to ideas about life. This, in turn, gives rise to various ethical, legal, and social issues. Here we discuss two key features in these issues: human dignity and human rights. To illustrate, we consider two significant areas in biotechnology: human genetic research and human embryonic stem (ES) cell research. In conclusion, we address economic implications.

HUMAN DIGNITY

Redefining Human Life

Progress in the life sciences and biotechnology has called into question what it means to be a human being. A couple, formerly diagnosed infertile, can now have a child with procreative assistance. An embryo, which in the past could only be examined after birth, can now be tested in the preimplantation stage so that genetic conditions can be detected, possibly eliminated or even selected. A young man, though apparently in good health, may be diagnosed through genetic testing as having Huntington's disease, but without the possibility of effective treatment. A child who lacks the enzyme adenosine deaminase (ADA), indispensable for survival, could be

25

treated under genetic therapy, yet the method is still uncertain. The genetic data for each individual may bring about individualized medicine, yet genetic particularities might also be used for discrimination. The clinical use of human embryonic stem cells to treat a variety of illnesses may soon be possible through regenerative medicine. However, this might arguably necessitate the use of somatic cloned embryos.

These are only a few examples of issues raised today by the life sciences. They show us how the distinctions between life and death, health and illness, normal and abnormal, just and unjust, and equality and discrimination are now vague. The more life sciences advance, the more significant is the question, what is a living human being? And from this fundamental question, come others: What is the essence of a human being's life? When does the life of a human being begin and end? And what does it mean to "live"? Although biotechnology brings us wonderful tools for health and medical care, it also confronts us with what appears to be the increasing materialization of human life. Because we cannot afford to neglect this, we need to inquire further about the value of human life, in particular, the meaning of human dignity.

Respect for Human Dignity

Human Dignity and Human Genetic Research

Human Dignity as a Universal Value   Since human genome research explores the mystery of human life in scientific fashion, core concepts about the human being are reconsidered. In this light, note the unique official global document, the Universal Declaration on the Human Genome and Human Rights, adopted by the United Nations Educational, Scientific, and Cultural Organization's (UNESCO) General Conference in November 1997 and endorsed by the United Nations' General Assembly in December 1998. At the beginning of the declaration, in section A, entitled "Human Dignity and the Human Genome," we read:

> Art.1: The human genome underlies the fundamental unity of all members of the human family, as well as the recognition of their inherent dignity and diversity. In a symbolic sense, it is the heritage of humanity.

This article points to the human genome as the basis for acknowledging human dignity. This means that the human genome cannot be subject to appropriation by any states, individuals, or other entities. Because the human genome represents the human species as a whole, research and application of the human genome should not be conducted if it harms human dignity. It should only be done for the benefit of present and future generations.

Article 2 goes on to state:

Art. 2: (a) Everyone has a right to respect for their dignity and for their rights regardless of their genetic characteristics. (b) That dignity makes it imperative not to reduce individuals to their genetic characteristics and to respect their uniqueness and diversity.

These paragraphs clearly highlight human dignity as the basis of human existence and thus of human rights.

Articles 6, 10, 11, 12, 15, 21, and 24 of the declaration continue to emphasize human dignity. Articles 11 and 24 are even more specific, both prohibiting applications that are contrary to human dignity. Article 11 states that practices that are contrary to human dignity, such as human reproductive cloning, should not be permitted. And in Article 24, the International Bioethics Committee identifies other practices that are contrary to human dignity, such as germ-line intervention.

"Human dignity" is a common term in philosophy and religion. It has often been used to differentiate between human beings and other entities like animals and plants. The term was actually introduced in legal language only after World War II, when the international community used the notion to determine crimes against humanity. Many international vehicles then adopted the idea as the fundamental value of human community. It is significant that "human dignity" has been used in conjunction with "human rights." While this chapter does not go into detail regarding the historical development of the idea of human dignity, what is relevant is its meaning within the context of bioscience and technology. In this respect, it is significant that the European Council on Biomedicine's Convention for the Protection of Human Rights and Dignity of the Human Being with Regard to the Application of Biology and Medicine uses human rights and human dignity together in its title. In Europe, particularly since 1990, human dignity has become a basic normative concept for the protection of human rights. Therefore, the meaning of human dignity is considered the basis for human rights.

The Meaning of Human Dignity in the Case of Human Reproductive Cloning   Human dignity remains a core concept in UNESCO's declaration, yet what does it mean? The idea seems self-evident, especially among Westerners. The term itself has not been subject to thorough examination, even during UNESCO's discussion of its draft declaration on the human genome and human rights mentioned previously. Yet for many other peoples, including Japanese, the concept of human dignity remains ambiguous.

For instance, note the recent discussion of human dignity by the Bioethics Committee of Japan in the context of prohibiting human reproductive cloning. After a lengthy and heated discussion on the issue, the committee determined the concept of "human dignity" with respect to human reproductive cloning entails three considerations: (1) human reproductive cloning instrumentalizes human beings for particular purposes, such organ transplantation; (2) predetermining genetic particularity is a violation of human individuality, that is, it violates respect for the individuality of each human person; (3) human reproductive cloning leads to family and social disorder in that it goes far beyond what is socially accepted as human conception. Taken together, these three considerations define human dignity, although in a negative way. They may also be applicable by way of analogy to other bioethical issues.

Human Dignity in the Case of Human Embryonic Stem Cell Research

Certainly one of the most difficult issues concerns research on human embryos, especially research on human embryonic stem cells. Embryonic stem cells (ES cells) have toti- or multipotentiality, unlimited differentiability, and the same number of chromosomes. They may be used for transplantation for the regeneration of cells and tissues. But such use requires the destruction of embryos at the blastocyst stage. Thus ES cells are derived from embryos through the destruction of embryos.

However, the embryo is an entity normally destined to become a human being. Thus the first question is, Is it permissible to destroy an embryo for research purposes? This is followed by a more general question: Is it ethically permissible to use an embryo for scientific research? If the answers to these questions are affirmative, then we need to explain the reasons as well as how this research can be regulated

Since an embryo represents an early stage in the life of a human being, we need to determine whether and from which stage an embryo can be recognized as a human being. From the moment of fertilization? After a period of seven or eight days, fourteen days, ten weeks, twenty-two or twenty-three weeks?[1] While it is clearly impossible to arrive at a unanimous answer, there are two responses. First, since life has already begun, no embryo should be the object of research. Using an embryo for any other purpose than birth is a violation of human life, and should be regarded as unethical and illegal. The other response is that, because the embryo cannot be deemed as a complete human being, since it is only a "thing," one may use the embryo for research. The nature of the embryo lies in-between these two extreme positions. That is, an embryo is surely not yet a human being, but neither is it merely a cluster of cells.

While there are various positions, it is not appropriate here to describe them all.[2] We may refer, for example, to the position taken by the Japanese Bioethics Committee in its review of the ethical issues in embryonic research, particularly human ES cells. The committee viewed the embryo as "a germ of human life."[3] Other countries use similar expressions, like "a potential human being" or "potentiality for human life."[4] In these cases, an embryo is neither a "thing" nor a "human being"; it is on its way to becoming a human being. An embryo thus is a being in-between and should be treated accordingly with respect to human dignity. The committee reported that creating a human embryo for the sole purpose of human reproduction should thereby be prohibited. The committee allowed the use of human embryos and human ES cells for research only in limited circumstances such as human-assisted reproduction or regenerative medicine. After the committee's recommendations, the Japanese Ministry of Education, Culture, Sports, Science, and Technology established guidelines on human ES cells research, which we consider in the following paragraphs.[5]

The use of embryos for human ES cell research is permitted in Japan. This gives way to a further question: From which embryos may such ES cells be taken? Four categories come to mind: (1) embryos created for the purpose of reproduction but turned over; (2) embryos specifically created for research; (3) supernumerary embryos; and (4) cloned embryos. The only permissible category is supernumerary embryos. This is because these embryos are already destined to be thrown out, that is, to lose their potentiality for human life. Therefore, it would be invaluable to use them for research that leads to effective treatments for incurable diseases. This, however, should not be thought of as a first step toward a wider and more general permission for embryo research. The principle remains the same—prohibiting the creation of embryos for research, yet using supernumerary embryos as an exception. Moreover, only a legally married couple may donate supernumerary embryos for human ES cell research. This restriction is consistent with the guidelines of the Japanese Association of Gynecologists and aims to avoid any easy donation of supernumerary embryos and thus to steer clear of the de facto "creation" of embryos for research. It should be noted however that in some other countries, like the United Kingdom, de jure marriage is not required.

Since human ES cells still have the possibility of becoming a human being on account of their totipotentiality, they should be treated with care even after they are derived. This means that human ES cells possess a special status, and they are not simply like other human cells or animal cells. Not only human embryos but also human ES cells "shall be handled carefully and consciously without violating human dignity" (Article 3 of the UNESCO declaration). More precisely, research institutions involved in human ES cell research should provide facilities specifically designed for human embryos or ES cells and distinct from facilities used for animal cells or tissues. Moreover, researchers dealing with human embryos or ES cells should be skilled and have ample experience with animal ES cells. Researchers lacking such skills and experience should not deal directly with human embryos and human ES cells.

The most difficult issue concerns so-called therapeutic cloning. The law that prohibits human reproductive cloning in Japan[6] does not ban the whole range of human cloning. Therapeutic cloning, or somatic cloned embryos, is regulated according to law-based guidelines. For now, however, therapeutic cloning is not permitted, pending further discussion on the status of the embryo by the Experts Investigation Panel of the Japanese Bioethics Committee. There is of course a risk of abusing therapeutic cloning. One major difference between therapeutic and reproductive cloning lies in whether or not the cloned embryo is introduced into the mother's uterus for gestation. Reproductive cloning is a criminal act with severe penalties, whereas therapeutic cloning is not forbidden by law. The critical question centers around determining the nature of therapeutic cloning. We acknowledge that the somatic nuclear transferred embryo is still a human embryo. Once we recognize that the cloned embryo is a human embryo, a special reason is still necessary to permit the creation of a cloned human embryo to be used for research.

As to research on human ES cells, it remains theoretically true that, once human ES cell research reaches the stage of clinical application, we will need to use ES cells derived from the nuclear-transferred embryo from the patient's somatic cell. Since there is no guarantee of success regarding ES cells research, there is reason to wait until clinical application becomes feasible enough. The Experts Investigation Panel, in its draft report on the status of the human embryo, is currently proposing a moratorium pending final decision after more years of appropriate human ES cell research development in Japan and worldwide.

THE PROTECTION OF HUMAN RIGHTS—
HUMAN GENOMIC RESEARCH

Human rights protection is critical in human genetic research. UNESCO's Universal Declaration on the Human Genome and Human Rights addresses the dialectic between freedom of research and the human rights of research subjects. Here, we limit our discussion to general principles in human genomic research. However, we should note that, in this post-genome-sequence era, after April 2003, we should pay more attention to issues concerning DNA data banking, where personal genetic data have become the source of future research and application. "Bio-bank" projects are already under way in countries such as Iceland, Estonia, Tonga, and Japan, while similar projects are starting up in other countries like the United Kingdom and Canada.

Freedom of Research

Freedom of scientific research is certainly one of the most cherished of freedoms. As discussed previously, because human genome research challenges views of human life, its applications also introduce ethical, legal, and social issues. Therefore, research should be carefully conducted in accordance with ethical and legal norms. This is particularly important when it comes to protecting human subjects who offer samples for genome research, without which such research cannot occur. From the viewpoint of the donors of the samples, to give their cells or tissues is an exercise of their rights over their bodies. Moreover, the outcome of genome research is genetic information, in turn the donor's personal information. So, those who give the samples should have their rights protected. In this dialectic of the freedom of research and the rights of the individual, two factors are crucial: informed consent and confidentiality of genetic information.

The UNESCO declaration addresses the need to protect these fundamental human rights. Article 12, paragraph b, clearly states that freedom of research, necessary for the progress of knowledge, is a fundamental feature in freedom of thought. The declaration also points out, in Article 13, inherent responsibilities of the researcher. Thus, no research is absolutely free. The core principles here remain the autonomy of the individual and the confidentiality of genetic data.

Informed Consent

The declaration requires that the individual who offers his or her bodily material as samples should give free and prior consent based on sufficient information. In many countries, any research that is conducted without informed consent is unethical as well as illegal. Article 5 provides a framework for this principle of autonomy. In particular, paragraph b stipulates three conditions of informed consent: prior, free, and informed. The latter means that the information should be conveyed in such a way that the individual clearly understands the meaning of the research, along with treatment and diagnosis.

Special attention should be paid to Article 5, paragraph d, designed to protect vulnerable persons such as minors, the mentally handicapped, and those suffering from dementia. The definition of "vulnerable" varies depending upon the situation, so that women as well as those who belong to minority groups may be vulnerable in various circumstances. There has already been some criticism of this article in that these conditions appear too rigid and unrealistic to be practically applied.

Confidentiality of Genetic Data

Genetic research on an individual reveals his or her genetic characteristics, which can be used for various purposes such as identification of disease, genetic vulnerability, and so forth. Therefore, the protection of human rights entails ensuring the confidentiality of genetic data of an identifiable person.

Article 5, paragraph c, also emphasizes that each individual has a right to decide whether or not to be informed of the results of genetic examination. According to this "right to know," researchers or medical doctors have the duty to inform the subject or patient about the results of genetic research or testing. From this is derived the "right not to know," since the genetic data taken from the sample is the personal data of the donor. Here the principle of the confidentiality also applies. Yet this raises a critical question: To whom does the genetic data belong? Because of its hereditary character, a good deal of genetic information is common among blood relatives. This evokes a further question: Does a relative have the right to know the genetic data of the donor of the sample? This is a very sensitive and difficult question, particularly if the donor has genetic traits of serious diseases like Huntington's or Alzheimer's. While we offer no definitive solution here, the U.S. National Bioethics Advisory Committee (NBAC) did report that it is not unethical to inform relatives of the results of genetic testing in cases of serious hereditary diseases. The Japanese Bioethics Committee concurred in its "Fundamental Principles Concerning the Research on the Human Genome."

Nondiscrimination

Genetic data can be abused for all sorts of reasons including employment discrimination, discrimination in insurance, and racial discrimination, even though, without any doubt, the principle of nondiscrimination is a core element in human rights protection. As the universal declaration clearly states in its Article 6:

> Art. 6: No one shall be subjected to discrimination based on genetic characteristics that is intended to infringe or has the effect of infringing human rights, fundamental freedoms and human dignity.

There are two types of discrimination: discrimination of a group and discrimination of an individual. The sharing of common genetic characteristics can possibly lead to the discrimination of groups such as ethnic or racial groups, minorities, and even groups of inhabitants in remote regions. Even if there is no difference in appearance, the results of genetic testing can reveal traits of certain groups and this can lead to discrimination. The same discrimination can occur for those with particular genetic traits for hereditary diseases.

Because genetic testing is simple and inexpensive, discrimination of the individual can easily result. Contexts for such discrimination include insurance, employment, marriage, and schooling. Insurance and employment discrimination reflects tension within the logic of market economy. Information regarding the genetic data of each employee or candidate for employment, or of the person insured or else seeking insurance, can be used against the person.[7] Of course, the significance of this insurance use depends on each country's own system of social security. In the United States, where the social security system is generally insufficient to provide a livelihood, health insurance assumes more importance. In some other countries, health insurance is supplementary to social security, and life insurance becomes that much more important. And note that when the government of the United Kingdom considered the possibility of insurance companies using the results of genetic testing in the case of Huntington's disease, the issue was so controversial that implementation has so far been suspended.

ETHICS AND THE PRINCIPLES OF MARKET ECONOMY

In all this, the tension between ethics and the logic of market economy is a crucial component. Economic factors have traditionally been perceived as irrelevant to ethical considerations. However, human genome research has far-reaching economic consequences, and these are indeed morally relevant. The first economic concern pertains to intellectual property. The second relates to the fact that human genome science and technology also raises socioeconomic issues such as aging or unequal medical care.

Intellectual Property Issues

Cutting-edge scientific research such as research on the human genome can bring high financial gain; it also depends upon sufficient financial resources. The fact of financial incentives as well as the recognition and prestige attributed to the "inventor" lie at the basis of the system of intellectual property. And even though the intellectual property system is currently dependent upon various national legislations, it should be clearly reaffirmed that no application of the human genome sequence should be patented. As former U.S. President Bill Clinton and U.K. Prime Minister Tony Blair clearly stated, "Raw fundamental data on the human genome, including the human DNA sequence and its variations, should be made freely available to scientists everywhere."[8] However, we cannot overlook the powerful impact of the intellectual property system. Clinton and Blair go on to state that "intellectual property protection for gene-based inventions will also play an important role in stimulating the development of important new health care products."

The intellectual property system is not an evil. It has two aspects. First, once a patent is admitted, all the information concerning this patent is open to the public. This publicity is one of the most important factors in the patenting system. Through this publicity, the patenting system serves to promote industrial development. Yet, unfortunately one can use patented information only after paying a royalty, a fee that may be too high for individual researchers. This, in turn, can disrupt their research. Here, major pharmaceutical enterprises enter the scene and, often in a monopolized fashion, buy the patent in order to produce new medical or pharmaceutical products. Thus, the final price of the product is considerably high in order to pay for future research and development.

Second, the issue of intellectual property is all the more important in the developing countries. This issue also concerns cooperation in transferring genetic technology and genetic data to developing countries. What is at stake still has to do with the publication of information and data that are collected in these countries.

We can neither deny the merits of the patent system nor the potential for its abuse out of economic gain. How then do we introduce ethical considerations when it comes to intellectual property in the case of human genome science? The line is hard to draw since there are also benefits for human health and welfare.

Social Costs Generated from the New Medicine and Pharmacology

We cannot overlook another point: economic implications of the application of the results of research. The development of human genome research has brought about highly advanced medical care such as genetic diagnosis and treatment. It has also produced a new scientific and industrial domain called "pharmacogenomics." And while such developments have contributed to maintaining and improving peoples' health, they have also generated other consequences.

The most important result is that the new pharmacogenomic products in medical care will certainly raise the costs of health care and pharmacological treatment. Surely, improved medical care can enhance the increasing population of elderly people. At the same time, this will impact the country's social security system. Herein lies the paradox: The more social security is assured, the higher the cost of the system. The burden of increasing numbers of elderly could become so taxing that, as a result, the social security system breaks down.

CONCLUSION

The themes presented in this article may appear contradictory. Cutting-edge science such as genome research needs to face certain social and ethical repercussions. What do these considerations mean for science? They make it clear that science cannot and should not stand alone. Indeed, science owes its existence to the society within which it is fashioned. Thus, society bestows a proper value upon science. This does not mean that science should simply be at the service of each individual society. In its broader sense, "society" refers to our global, human community, where human values are respected. Thus, freedom of research, a fundamental human right, should not displace all other human values. The major question concerns whether or not scientific research is conducted in ways that respect human values. Recognizing the social dimensions of science and scientific progress helps to ensure that we maintain the equilibrium between science as a promoter of a given society's abilities and public awareness of the applications of science for the welfare of that society. Thus the core of this reflection lies in harmony. Thirty years ago, the main theme of the Universal Exposition in Osaka was "Human Progress in Harmony." We now need to revive this same theme for the twenty-first century.

NOTES

1. The International Bioethics Committee of UNESCO highlighted the differences among various religions concerning this question in its insightful report on "The Use of Embryonic Stem Cells in Therapeutic Research," BIO-7/00GT-1/2(Rev.3), 6–7.

2. Refer to the International Bioethics Committee of UNESCO report on "The Use of Embryonic Stem Cells in Therapeutic Research" for a full ethical analysis.

3. Japanese Bioethics Committee, Science and Technology Council, *Basic Conception on the Research on Human Embryos, Focusing on the Human Embryonic Stem Cells* (in Japanese), March 2000, 8.

4. The Singapore Bioethics Advisory Commission determined that embryo has a "special status," and concluded that the use of the human embryo for human ES cell research should be permitted. It should be noted that it is possible in Singapore to donate an embryo for ES cell research abroad.

5. Japan Ministry of Education, Culture, Sports, Science and Technology, *Guidelines on the Derivation and Utilization of Human Embryonic Stem Cells*, 2001, English version is available at www.mext.go.jp/a_menu/shinkou/seimei/index.htm.

6. Law of November 30, 2000, concerning regulation relating to the techniques of human reproductive cloning and other similar techniques, www.mext.go.jp/a_menu/shinkou/seimei/eclone.pdf.

7. Former President Clinton's statement on banning discrimination of government officials in this regard clearly indicates the gravity of this question.

8. D. Evans, "Clinton and Blair Want Free Access to Genes Map," London (Reuters); www.mansfieldet.org/schools/mms/staff/hand/genartclinton.htm (accessed March 14, 2000).

II

SPECIFIC CHALLENGES IN CULTURES AND NATIONS

Now we examine specific issues raised within specific countries and cultures. Though generated by what Chalmers calls the "scientific power blocks" mainly in the West, biotechnology's applications extend throughout the world. They extend in ways that are intended to meet the needs of specific societies and cultures. Yet bear in mind that perspectives on biotechnologies are not universally shared. These perspectives reflect fundamental values within specific cultures. The challenge lies in determining whether or not these values have common ground. Can we cultivate a shared vision with respect to biotechnology?

The link between genetic research and reproductive technologies has become an area where the United States remains pretty much in the forefront of research. In this context, Jeffrey Kahn and Anna Mastroianni address a complex and controversial application of preimplantation genetic diagnosis (PGD). They explore the much-publicized Molly Nash case, in which PGD was used to ascertain a nondisease trait for Molly who was in need of a hematopoetic stem cell transplant because she was born with Fanconi anemia, of which both parents were carriers. The authors deftly take us through the drama, the ethical issues and debate, and current problems with U.S. policy. They conclude with constructive suggestions for enhancing ethical sensitivity as well as establishing a more just public policy.

It becomes apparent that Kahn's own involvement with the Nash case was in-depth and sensitive. His engagement, however, did not diminish his ability to squarely address key ethical concerns such as questions regarding parental motivation. Essentially, the fundamental moral question that the authors bring to the table deals with limits. Should there be moral limits to such uses of medical technology? How do we fairly resolve the matter of

37

limits when it conflicts with notions of procreative freedom, of particular concern in much of U.S. culture? Kahn and Mastroianni astutely point out policy implications, current policy weaknesses, and provide recommendations, such as recommendations centering on the need for sufficiently sound, well-informed discussion and dialogue.

Just as PGD presents policy challenges in the United States, genetic testing in Canada also poses its share of regulatory concerns. Mylène Deschênes ably discusses the unique regulatory framework of our northern neighbors regarding genetic testing. After exploring how public perceptions affect public policy, she gives us a clear look at questions pertaining to the quality and applications of genetic tests. In her critical assessment of the Canadian regulatory framework, she cites its "patchwork" approach and underscores areas of need.

Deschênes acknowledges that Canada can learn from the U.S. regulatory framework. After a discussion of relevant policies and documents in both the United States and the United Kingdom, she then offers ways to improve Canada's current system. She notes that Canada is in a particularly advantageous position due to its unique system of universal health care, a system that can bring about the greatest health benefits for Canadians when it comes to genetic testing. Furthermore, high quality standards are not compromised in Canada's system. Hers is a critical voice reminding us of the need to balance access and safety. Moreover, a risk–benefit assessment cannot afford to be solely defined within scientific parameters. Social, cultural, and ethical considerations enter into the equation as well.

The debate rages on in the United States over the morality of experimentation with human embryos. What about other countries? Heinrich Ganthaler nicely discusses the debate in Europe. Despite embryonic research's promise of enormous benefits in alleviating debilitating disorders, European legislation generally remains cautious and restrictive. At the same time, there is no unanimity among European countries.

Ganthaler discusses this European debate in sound fashion. Yet his contribution is especially valuable in his careful summary of the official positions and statements in Austria and Germany by presenting an enlightening review of their respective committees on bioethics. He then goes on to examine in more detail the philosophical arguments both for and against human embryo research. Ganthaler concludes that the benefits to actual persons outweigh the harms to potential persons (human embryos) and, given the absence of alternatives, such research is ethically justified.

How has biotechnology influenced the cluster of countries in the former Soviet republic, the states acceding to the European Union, and the Balkans? Larissa Zhiganova and Yuri Gariev refer to these countries as "post-Communist Europe," and contend that biotechnologies continue to make a definite impact upon this entire region. At the same time, they incisively point out how the interplay between biotechnology and culture produces its own unique dynamic. This dynamic manifests cultural values that need to be considered in order to understand the impact of biotechnology in post-Communist Europe.

The authors first examine the past cultural context, including the force of Marxist-Leninist ideology, Russian statism, and religious nihilism as providing contexts for revolutionary discoveries in sciences. They then expose the current cultural context and the challenges posed by biotechnologies, such as the absence of emphasis upon a free-market framework, and disruptions of historic continuity. Nevertheless, studies in biology, genetics, and bioethics continue to assume increasing significance. Zhiganova and Gariev conclude by discussing future prospects in view of major challenges to overcome a Communist legacy. At the same time, they point out the critical need to redress current distortions in marketing certain products. In all of this, open discussion and dialogue, listening to all of the voices at stake, is imperative. Cultural sensitivity is the key.

What about Jewish perspectives on biotechnology? As much as we emphasize the need for cross-cultural conversation and understanding, we find few analyses of the rich tradition of Judaism, a tradition with a long history and complex culture. Edward Reichman faces the challenge expertly. He covers a host of topics in biotechnology including genetic testing, preimplantation genetic diagnosis, stem cell research, and human cloning.

Throughout his discussion, Reichman offers us an invaluable look at Orthodox Judaism, its scriptural and codified bases, its nuances, and its flexibility with respect to certain applications of biotechnologies. He skillfully draws upon biblical and Talmudic sources. In his analysis, it is clear that Judaism continues to be a prolific and dynamic tradition that, while adaptable to medical and biotechnological advances, remains true to its foundations.

The same dearth of understanding and analysis with respect to Judaism is even more so regarding the long-standing tradition and culture of Islam. Bushra Mirza provides an invaluable summary and discussion of the most prominent Islamic beliefs that are relevant to biotechnology. These days, more so than ever, we need to know and understand more of the Islamic tradition. Her discussion, clear and crisp, offers an excellent introduction to Muslim views on biotechnology.

Grounded upon the authority of the Quran and Sharia (Islamic law), the key Muslim values of respect for human life, human community, and justice underscore Muslim views towards biotechnology. This clearly prohibits exploitation in any form. This has direct relevance for matters such as genetically modified foods, reproductive technologies, genetic screening, embryo research, gene therapy, and human cloning. At the same time, Mirza also points out guiding principles that demonstrate a flexibility in Islamic law. Moreover, some issues still remain open to dispute among Islamic scholars.

Nevertheless, Islamic scholars all agree that biotechnologies must be utilized to alleviate human suffering.

With 13 percent of the world's population, increasing food demands, and diminishing environmental sustainability, Africa is the test of biotechnology's risk–benefit ratio, particularly when it comes to agricultural application. Conventional technologies cannot keep up with food demands and an increasing population. Martin Makinde poignantly examines the critical issue of hunger, particularly in sub-Saharan Africa, and claims that biotechnology holds great promise to ensure better food security for the continent. He cites persuasive reasons why genetically modified crops can produce numerous and far-reaching benefits.

Despite this enormous potential for enhancing development, however, there are concerns that need to be addressed, having to do with biosafety, health, industry, environment, policy, and the potential for exploitation when it comes to patenting and intellectual property rights. To deal with these challenges, Africa needs to establish its own regulatory mechanisms in addition to international agreements and partnerships. At the same time, there needs to be a full-fledged educational effort to counter negative public perceptions. According to Makinde, unless the issues of equity are addressed, there will be dire consequences for all peoples, not only for those in the developing world. His is a clarion call to establish proper priorities and sound mechanisms.

David Chan poses some rather hard-hitting questions when it comes to certain assumptions the West may have about so-called fundamental principles. When medical research is conducted in East Asian countries with a rich tradition of Confucianism, does the principle of autonomy still occupy a leading role? If not, should these countries adopt Western principles? How consistent are these principles with Confucian values? Similar questions can be posed regarding the application of biotechnologies.

In his thoughtful review of ancient Chinese medical ethics, Chan reminds us that Confucianism stresses the values of humaneness and compassion. He then highlights the tensions between Western principles and deep-rooted cultural values through a recent case in Singapore of research ethics gone awry. Chan concludes that relying solely on the traditional principles of beneficence and nonmaleficence is not enough. To ensure more ethical research, autonomy and informed consent must be implemented. At the same time, this must be complemented by sensitivity to Confucian values. He claims that respect for patient autonomy can indeed legitimately cross cultures without violating Confucian values. Singapore is an example of a Chinese country moving in the direction of acknowledging the principle of autonomy.

Notions of intellectual property rights hinge upon assumptions that lie imbedded in what Stella Gonzalez-Arnal calls the "knowledge economy." Along these lines, in discerning fashion, Gonzalez-Arnal examines the claim that many international agreements regarding intellectual property rights are in essence skewed. That is, they are biased in that they make assumptions based upon a rather narrow epistemic model, or knowledge paradigm. This model favors views of knowledge that may be the norm for much of Western thinking. However, it is not necessarily the norm for more traditional and indigenous cultures. Gonzalez-Arnal contends that assuming the universality of this narrow model through the current process of patenting can threaten the knowledge base and lifestyles of these indigenous communities.

Gonzalez-Arnal further claims that there is definitely an epistemic value to traditional knowledge systems. Furthermore, there is in essence a plurality of knowledge systems. The challenge lies in establishing tools of intellectual property law that can protect and sustain this plurality. She rightly inquires whether and how we can fairly reward the intellectual labor of communities. In her insightful philosophical analysis, acknowledging and embracing this will in effect benefit everyone.

4

The Ethics and Policy Issues in Creating a Stem Cell Donor: A Case Study in Reproductive Genetics[1]

Jeffrey P. Kahn and Anna C. Mastroianni

During the nearly ten years of its existence, preimplantation genetic diagnosis (PGD) has been used predominantly to avoid giving birth to a child with identified genetic disease. Recently, PGD was used by a couple not only to test embryos created by in vitro fertilization (IVF) for genetic disease, but also to test for a nondisease trait related to immune compatibility with a child in the family in need of a hematopoetic stem cell transplant. This chapter describes the case, analyzes some of the ethical issues it raises, highlights gaps in U.S. policy, and, finally, makes some ethics and policy recommendations for addressing advancing genetic and reproductive technologies.

THE NASH FAMILY CASE

The story of the Nash family and their successful use of preimplantation genetic diagnosis to cure their daughter received national attention for a variety of reasons.[2] First, it is a compelling human interest story, and one with a happy ending. Second, for many in the national media it raised the specter of genetic testing run amok. Third, and most importantly for this discussion, it is a case that raises numerous ethical issues and exposes the lack of institutional or policy controls over the burgeoning uses of genetic and stem cell technologies. In this chapter, we examine the issues raised by the case, and make some recommendations regarding the growing need for oversight mechanisms.

The Nash family relied on the relatively new technology of PGD, which allows for genetic testing of very early-stage human embryos prior to their implantation. PGD relies on traditional IVF techniques, followed by a "biopsy" of the embryo at the eight-cell stage at two to three days postfertilization. The biopsy is performed by nicking the embryo's outer membrane, and then removing one of the eight dividing cells. The DNA is removed from the single removed cell, and genetic testing techniques are applied to the DNA. PGD has been used to help prospective parents avoid bearing children with genetic diseases, primarily in cases where parents have known genetic risks.

In the Nash case, both parents were carriers for the genetic disease Fanconi anemia (FA) but were unaware of their carrier status. Since FA is a recessive genetic disease, when two carriers mate there is a one in four chance that the disease will affect their offspring. In the Nashes' case, their first child, Molly, was born with FA after being conceived the "normal" way, that is, without medical intervention. Children born with FA face a number of obstacles. There are physical problems typical to children with FA, including fused joints in the hips and wrists, missing thumbs, and incomplete guts and, most important for this story, the children become leukemic at six- to eight-years old (Wagner et al. 1999). To treat the leukemia associated with FA, the children require a hematopoetic stem cell (HSC) transplant for their survival. HSC can be donated either by collection of bone marrow or peripheral blood stem cells or collected from the umbilical cord blood after a baby's birth. The most likely HSC donors are siblings, since they are the closest genetic relatives to the patient. In the Nash case, Molly was their first and only child, so no sibling donors were available. The next best donor source would have been other relatives, but nobody in the family was a sufficiently close immune match to qualify as a donor. When no related donors are available, the National Marrow Donor Program[3] can be used to match unrelated individuals willing to be donors with patients in need of an HSC transplant.

Research has shown that in children with FA, those with sibling transplants have a substantially higher success rate than those with transplants from unrelated donors (Wagner et al. 1999), so the Nashes hoped that a future child would be closely enough matched to act as a donor of umbilical cord blood stem cells. Other couples who found themselves in similar situations had gone through a process of deliberate conception followed by prenatal diagnosis to determine whether the fetus was (1) FA negative and (2) immune-matched to their sick child. The results of the prenatal testing led to a potential abortion decision if the developing fetus was found to be carrying the FA mutation. In at least two documented cases, couples aborted otherwise healthy fetuses that were FA negative but immune incompatible with their sick children (Auerbach 1994). Because of the limited time in which to find a donor for Molly and to avoid the need for an abortion decision, the Nashes sought to use PGD to test embryos made in vitro rather than going through the process of prenatal testing and subsequent decisions about abortion. The PGD is a two-stage process: testing first for the FA mutation, and then testing for human leukocyte antigen (HLA) compatibility with Molly among those embryos that tested FA-negative. Five separate times over a period of many months, the Nashes went through the process of collecting ova and IVF in Denver followed by PGD performed by a lab in Chicago before achieving a successful pregnancy with an FA-negative, HLA-matched embryo. That pregnancy resulted in the birth of Adam—so named for the biblical story that it was one of Adam's ribs that was used to create Eve—in August 2000 in Denver. The umbilical cord blood was collected and flown to Minneapolis where it was frozen until the hematopoetic stem cells it contained were infused into Molly in September 2000 at the University of Minnesota. One hundred days later a news conference was held at the university, where it was announced that Molly's bone marrow was identical to that of her brother Adam, evidence that the transplant had been successful. The Nashes returned to their home in Denver in January 2001, where Molly for the first time enjoyed life as a healthy child.

Why Does This Case Matter?

This case raises a range of ethical and policy issues, and serves as a very effective example for both the types of social issues that we face as biotechnology advances, and challenges us to consider whether there are limits of ethical acceptability and, if so, what they ought to be. It is a useful case in that it is far from hypothetical, with real people whose names and story make them seem like us—forcing us to consider what we would do faced with similar choices. The fact that the Nashes were able to use a combination of existing technologies in the way they did can also serve as an object lesson for why we need to think about institutional and policy controls. Their case offers a glimpse into some of the real ways that stem cell technologies will be used. Finally, the Nash case serves as a concrete example around which to craft principles, rules, or frameworks. If such approaches prove sufficiently robust, they can help inform approaches for addressing other controversial policy areas such as other reproductive technologies, stem cell research issues, and even cloning.

SOME ETHICAL ISSUES

The Nash case is interesting as both a human interest story and for the ethical and policy issues it raises. In the publicity surrounding the case, many pundits questioned both what characteristics the Nashes chose through the use of PGD and their motivations for choosing them (Belkin 2001). Examining these general claims is one way of exploring the ethical and policy issues the case raises.

Do Characteristics Chosen Matter?

Some of the concern raised by the Nashes' behavior centered on the mistaken claim that they had manipulated human embryos to create a stem cell donor for their daughter. This is a misunderstanding of the function of PGD, of course, which can only be used to test, not alter, the genetic makeup of an existing embryo. Any sort of genetic manipulation would be on the order of gene therapy, a technology yet to show successful application aside from a very few patients. What had happened in the Nashes' use of PGD, however, was that a line had been crossed. In the seven or so years since PGD has been developed and introduced, its use has been restricted to avoiding disease in future children for couples at risk of passing on genetic disease. The most common use has been to avoid bearing children with diseases like cystic fibrosis, but the technology can be used to test for any disease (or other trait) for which a genetic test has been developed.

In the Nash case, however, this distinction had been ignored. The first stage of PGD was in fact used to avoid disease—embryos were screened and only those that were Fanconi-negative proceeded to the second stage of testing. The second stage crossed the line between avoiding genetic disease and selecting for some nondisease trait by testing for HLA status. Further complicating the analysis, the selection of HLA status was not to benefit the child that would develop from the tested embryo, but to ensure immune compatibility with the future child's sister Molly. So not only was PGD used to select for a nondisease trait, it was used to select a nondisease trait selected to benefit somebody other than the child who would be born. For some, this is softened by the argument that were it not for the fact that Adam was both FA-negative and HLA-matched to his sister, he would not have been born. On this argument, the choice of a nondisease trait to benefit Molly actually also carried the ultimate benefit to Adam of being brought into the world (Robertson 1994; Robertson et al. 2002; Parfit 1984). The problem with such an argument is that it can justify selecting nearly any characteristic one chooses in the embryo on the grounds that it is better to be brought into the world under such circumstances than not to exist at all. We argue below that such arguments ought to have their limits.

As genetic research yields increasing information about both disease and nondisease traits, it is only a matter of time until couples choose not only disease-free embryos, but embryos that have particular physical or behavioral characteristics, such as musical aptitude, athletic ability, outgoing personality, blue eyes, above average height, and so on. The only limits in sight seem to be what tests are available, which will only increase as the Human Genome Project yields more and more meaningful results. Which characteristics are chosen, or more correctly, which characteristics are avoided, are so far up to the couple and their health-care provider to decide. But as more tests become available, we should be increasingly uncomfortable leaving decisions that have both individual and societal implications up to individuals alone to decide. At the very least, we ought to give some guidance to parents and physicians about the limits of acceptable uses of genetic testing of embryos.

One basic standard should be that the characteristics chosen be in the best interests of the child who will be born, not merely to benefit someone else, such as parents or siblings. In the Nash case, a particular HLA status was selected to help save Molly's life rather than to directly benefit the child that is Adam. However, having one combination of HLA antigens versus another has no effect one way or the other on Adam's health, so its selection is effectively neutral for him. Given that testing was also used to ensure that the embryos selected (including Adam) were negative for Fanconi anemia, the best interest standard was met in the Nashes' selection of a Fanconi negative and HLA-matched embryo that would produce baby Adam. This conclusion forces a more refined analysis of whether the couple's motivations for having a donor child matters.

Does Motivation Matter?

A consistent claim in discussions about the Nash case was that their motivation for having Adam was suspect in that he was brought into the world at least partly to save the life of his sister. The concern in these claims is that parents could use predictive genetic testing technologies like PGD to serve motivations that served themselves or their existing children but had very little to do with the interests of the future child. But why should the Nashes' motivation be open to this sort of second-guessing and assessment of moral propriety and not those of other parents? Were we to have access to the true reasons that people have children, we'd find everything from the "right" answer of the desire to bring children into the world for the intrinsic value they have, to love and cherish them, and nurture them in caring environments, to less wholesome motivations like carrying on one's family legacy and having siblings for one's other children. In fact, if we understood the true reasons behind many children's births we'd hear that they were accidents, unplanned pregnancies due to failed birth control or just plain carelessness. In many cultures, including some agricultural areas of the United States, families historically have had many children in order to ensure that there are enough hands to do the work required and to care for the other children in the family. In short, we need to think hard about whether there are wrong reasons for having children, and, if there are, what we might do about it. Given the fact of the very wide range of reasons and motivations for having children, it is difficult to convincingly argue that having a child to save the life of an existing sick child is such a bad answer to the question.

All this being said, there are still limits we impose on what parents do with the children that they bear, and such limits may be instructive for limits in using PGD. Parents are prevented from abusing or neglecting their children, with the state stepping in and even removing children from their parents when their health and safety are threatened. We can unfortunately envision cases when parents create children to serve their own or their other children's interests in ways that could violate those limits. From a moral perspective, we want to prevent parents from violating the Kantian norm of treating their children as ends unto themselves and never as a mere means to the ends of others. Consider two examples that arguably cross that line. For one, the Nashes could have gone through the process of IVF, followed by PGD to select an embryo that was both FA-negative and HLA compatible with Molly, eventually resulting in a baby whose umbilical cord blood could be collected and used as a transplant for Molly. But instead of taking Adam home, the Nashes could have put him up for adoption. Instead of bringing Adam into the world to love and cherish as their child, the couple would have effectively brought Adam into the world for the cells in his umbilical cord—not for him, but for his parts. This seems to be using Adam as a mere means to his parents' and Molly's ends rather than treating him as an end unto himself.

A second scenario is less clear. It turns out that some couples have figured out that the same cells that will be in the umbilical cord blood at birth are in the fetal liver after approximately sixteen weeks of development. So rather than wait for the baby to be born, the fetus could be aborted after sixteen weeks and the hematopoetic stem cells collected from the fetus's liver. Rather than being a speculative scenario, there are reports of a few couples who have asked to pursue this approach in cases where they have a son with adrenoleukodystrophy (ALD, or Lorenzo's Oil disease), which can be treated by HSC transplants (Boyce 2003). Only boys are affected by ALD since it is an X-chromosome-linked trait. But heterozygous girls have an increased risk of other health problems, meaning the only truly "normal" children are those without the trait. What this means is that the odds of finding an unaffected embryo that is also HLA-matched are much lower, so the parents are willing to implant heterozygous female embryos with the intent of aborting them after sixteen weeks. This would avoid bearing a child with the health risks associated with ALD trait, but allow the collection of potentially lifesaving hematopoetic stem cells from the liver of the aborted fetus. Whether this example qualifies as a mere means use depends on how we understand the status of the human fetus—the prohibition is on the mere means use of another *person*, and for many a human fetus does not qualify. Whatever the answer, it is a worrisome behavior that ought to be prevented if possible. In fact, the public law that allows the use of federal funding for fetal tissue research also bars the directed donation and use of discarded fetal tissue, making the scenario described a criminal offense.

Given this context and the realities of parents like the Nashes faced with ill children, what moral principles can we propose to protect future children while at the same time preventing the misuse of children "created" not as ends in themselves but as means to the ends of others? At the very least, we argue first for the selection of characteristics that are in the best interests of the child who will be born, and second that the treatment of the child after he or she is born be limited in terms of the physical risks posed. To put it bluntly, we would hope to avoid the situation of couples literally creating children for the parts they can provide. It will be very difficult to prevent the case of a couple conceiving a child with the intention to put it up for adoption without something akin to licenses for parents, which are both an unacceptable infringement on procreative liberty and an impractical if not impossible to enforce option. More successful will be efforts to oversee the treatment of the children after they are born, and make sure that appropriate risk–benefit balance exists when children are used as donors. This is not an idle concern, since HLA-matched children can donate not only umbilical cord blood (which has no risk) but, in the event of a failed cord blood transplant, could also be used as bone marrow donors, which carries much greater risk of morbidity and even mortality. It is even expected that children treated by HSC transplants for Fanconi anemia will eventually likely need kidney transplants due to long-term use of immunosuppressive drugs, and their HLA-matched siblings will again be obvious potential donors. Since parents are in a position of conflict in deciding whether an HLA-matched child ought to be a donor for his or her sibling, third-party review offers a mechanism for ensuring that prospective donors are not exposed to greater than acceptable risks for the benefit of their siblings. Such review could assess whether there is sufficient medical and psychological benefit to the donor to offset the risks inherent in the donation.

POLICY IMPLICATIONS

What makes the Nash case and its implications so challenging? The case used a new combination of existing technologies—creation of embryos by IVF, use of PGD for selection of traits, and collection and use of umbilical cord blood for transplant. Each technology alone has been the subject of ethical debate and policy making; but when they are used in combination, the discussions become more complex and ultimately expose a policy gap. Put another way, the combination of technologies falls between the cracks of existing policy approaches for determining appropriate uses and controls of controversial medical technologies.

This policy gap exposed by the Nash case exists because there are few, if any, mechanisms for assessing the acceptable uses of each technique employed in the case and a dispersion of responsibility for making such assessments. We have identified three components of this policy gap, each discussed in turn: (1) multiple sites leading to no locus of overall responsibility; (2) limited mechanisms for assessing acceptable creation and uses of human embryos; and (3) limited third-party oversight of the medical technologies involved.

No Locus of Overall Responsibility

The Nash case makes clear that whatever controls we might suggest, the fact that the various elements of the process can take place at different sites makes oversight difficult. In the Nash case, IVF was performed at a clinic in Denver, PGD in Chicago, and the cord blood transplant in Minneapolis. Rules or oversight dictated by the IVF clinic have little impact on behavior at the PGD clinic or in the transplant unit, and vice versa. The upshot of having multiple sites for the individual elements is that there is no locus of overall responsibility for the process, and creates an environment in which each of the individuals and institutions involved can claim that the implications are out of their control.

No Mechanisms for Assessing Creation and Uses of Embryos

There are few, if any, mechanisms for assessing the acceptable creation and uses of human embryos, particularly in the medical context. Part of the policy gap is related to the practice of reproductive medicine and the creation of human embryos. Since the vast majority of embryos created in the United States and abroad are the product of IVF, one potential area for control would be on the reproductive medicine clinics that perform IVF. Recent surveys suggest that there are more than four hundred thousand leftover embryos in the United States alone (Hoffman et al. 2003), and as numerous others have pointed out (Andrews 2000; Knowles 2002), reproductive medicine is among the least regulated or controlled areas of medicine.

In 1981, during the first Reagan administration, a ban, which continues to the present, was imposed on the use of federal funds for any research that harms or destroys human embryos (Public Law No. 105-277, 1998). While a research ban would seem to provide clear policy direction, it has in fact created a laissez-faire policy environment, or policy gap, related to embryo research. This may seem counterintuitive given the embryo research ban in place for more than twenty years, but it is precisely because of the ban that so few controls exist. When the federal government agrees to fund a particular program or area of research, the funding always comes with strings attached. In effect, the government says that if you want its money, then you must agree to follow its rules. This can be seen throughout government programs, from educational programs to large transportation projects to clinical trials. However, when there is no federal funding there are no funding-related rules or restrictions. This is the case with embryo research, since the government's policy only bars federal funds being used for embryo research and therefore embryos can be created, destroyed, experimented upon, and used for any purpose so long as no federal dollars are used. This has left embryo research as one of the few areas of biomedicine that is carried out exclusively in the private sector and effectively unregulated.

Limited Third-party Oversight of the Technologies Involved

Finally, there is very little third-party oversight of reproductive medicine by payers, the government, or professional organizations. Reproductive medicine has long enjoyed a market-oriented approach to oversight, in part because such a large proportion of the costs of reproductive medicine services are borne by patients directly and not by third-party payers. Since insurers pay so little of the costs of IVF and other reproductive medicine services, they have little say over the appropriate uses of the technologies involved. Instead, it is left to market forces to decide what restrictions ought to exist, if any. Likewise any attempts at self-regulation by reproductive medicine specialists are more influenced by what patients demand and are willing to pay for than by what the profession might deem as appropriate.

To summarize, there is a policy gap (which could easily occur again) when it comes to attempts to control the efforts to create immune-matched stem cell donors. This stems from the combination of multiple sites of responsibility and lack of any locus of responsibility for the overall process; the limited oversight of IVF and other reproductive medicine services owing in part to the embryo research ban; and the market-driven nature of reproductive medicine. This policy gap has implications not only for use of the individual technologies involved, but for how they might be used in cases like the Nashes and others in the future. Without reflective policy making, we are more likely to see policy made by reaction to scandalous cases and the "yuck factor" associated with them.

Some Implications

The policy gap revealed by attempts to create HLA-matched donors highlights concerns about the extent to which couples may use genetic testing to identify traits in their future children. Many of the media reports on the Nash case charged them with creating a "designer baby." While the selection of the traits identified in a range of embryos is not the product of genetic manipulation or other technologies more akin to designing, the Nashes did select from the genetic testing menu available at the time. That menu will only become larger and more detailed as genetic information and tests created from it proliferate. It is clear that some parents will use whatever tests are available, including tests for physical and behavioral characteristics. In fact, reports that some reproductive medicine clinics are offering and some couples are using PGD to select for gender is evidence that the use of genetic testing is limited by what is available rather than by what couples will use.

While the Nash case highlighted a two-stage process of genetic testing (testing for the FA mutation, followed by testing for HLA status), there are likely to be many people who test embryos for HLA status when there is no risk for genetic disease. Anybody with a disease that could be treated by an HSC transplant and could wait the nine months it will take for a matched donor to be born could use PGD to create a donor. This would apply to adult and childhood leukemia, rheumatoid arthritis, and numerous other diseases, greatly increasing the potential demand for combined technologies.

An increase in demand would create huge cost and access issues. The Nashes spent more than $100,000 for five attempts of both IVF and PGD before achieving a successful pregnancy. These costs did not include the cost of any treatment for Molly, and all the costs (IVF, PGD, and stem cell transplant) were borne by the family since their insurance refused to cover what they deemed to be experimental treatment. By their own admission, the Nashes were very fortunate to be able to afford the extraordinary costs of treating their daughter, but the vast majority of families would not be so lucky. If we conclude that the use of PGD to create stem cell donors is acceptable as a matter of policy, how then are we to ensure equitable access to the technologies? In the United States, we often end up paying for high-priced life-saving therapies when they are brought to the public's attention. This ad hoc sort of resource allocation can work when there are only a few cases to address, but is irresponsible because it relies on no reflective process and is a poor way to spend increasingly limited health-care resources.

Change may be coming, however, in that third-party payers may find it increasingly difficult to deny coverage for the creation of stem cell donors for at least two reasons. First, the technologies being used—IVF, PGD and umbilical cord blood transplant—are all part of mainstream medical care. Second, many insurers deny coverage for IVF on the grounds (rightly or wrongly) that infertility is not an illness or disease. But in the case of creating a stem cell donor, IVF is not being used to treat infertility (the Nashes were not infertile) but as part of a treatment for the HSC transplant recipient, effectively creating a therapeutic use of IVF. The first appeals to insurers have been made on these grounds, and no doubt lawsuits will follow.

At least two lawsuits have already been brought (though unsuccessful) against physicians on the grounds that they failed to inform families of the option of using PGD to create stem cell donors that could have saved the lives of their sick children. It is only a matter of time before similar suits are brought against genetic counselors and others, with the associated costs they will bring.

Finally, attempts to create HLA-matched stem cell donors have the unintended consequence of creating a potentially large number of leftover or "spare" human embryos. First, couples who would otherwise not avail themselves of reproductive medicine services use IVF to create embryos that can be tested by PGD. Second, these couples create far more embryos than in "regular" uses of IVF since it takes a relatively large number of embryos to get even one that is both disease-negative and HLA-matched. The Nashes reported that they created upward of twenty-five embryos, many of which remain frozen. Other couples who unsuccessfully attempted to create a matched donor have created upward of one hundred leftover embryos. While this problem is not unique to the creation of stem cell donors, it highlights the fact that there is no policy for the disposition of leftover embryos in the United States—where recent reports placed the number of frozen embryos at upward of four hundred thousand (Hoffman et al. 2003).

SOME POLICY RECOMMENDATIONS

How can we improve the policy environment for intervention and oversight in the creation of stem cell donors? A few concrete recommendations follow:

(1) Move the debate from the clinic to the public policy arena. Ethical and policy issues are currently addressed by individual physicians in discussion with individual patients. While this is an appropriate model for much of health care, the making and testing of human embryos has societal as well as personal implications, and the societal issues should be addressed as matters of public policy. There are many potential forums for such discussion and debate, including the Institute of Medicine (IOM), National Institutes of Health (NIH), and others.

(2) Avoid reactive policy making. One of the well-founded worries in the medical community is that a very provocative and controversial case will make its way into the news media and result in reactive and potentially knee-jerk policy making. The only way to avoid this is through organized public policy debate and institutional commitments to local review until wider policies are promulgated.

(3) Create local mechanisms for review and advice of controversial uses of biomedical technologies. Institutions can do their part by establishing processes for at least advisory review of new and controversial applications of biomedical technologies. One example is the University of Minnesota's Stem Cell Ethics Advisory Board,[4] which exists to provide advice to any university faculty or staff involved in the use of stem cell technologies. The

board's membership is a combination of internal and external experts in medicine, science, law, ethics, and religion, and is linked to the university's internal review board (IRB) by ex officio membership of the IRB executive chair. The board's conclusions and recommendations are advisory but not binding, and the university strongly encourages but does not require that stem cell researchers consult the board.

(4) Consider lessons from others. There are policy models in other countries that can be instructive for moving forward in this discussion. The United Kingdom's Human Fertilisation and Embryology Authority (HFEA) offers strong central control of any creation and use of human embryos. In so doing, it can limit the uses of embryo creation and testing as the authority rules and can review requests for uses that fall outside of existing policy. The same mechanism allows for more liberal research uses of embryos than U.S. policy. Such a centralized approach would require a wholesale shift in not only embryo research oversight but also the structure of reproductive medicine in the United States, making this approach very unlikely. What is instructive is that while centralized control is most often viewed as a restrictive approach to oversight, in practice it can actually allow for more liberal policies than diffuse controls.

CONCLUSION

It is clear that cases like the Nashes' raise significant and challenging ethical and policy issues, and that we lack sufficient mechanisms for addressing them. The case pits fundamental core principles such as procreative liberty and prohibition of mere means uses of individuals against each other, so it is no wonder that the issues are so difficult to address. But the issues won't go away, and we must address them since the beneficial uses of these medical technologies are at stake. Reactive policy making is not the answer, nor is unchecked uses of existing and new technologies. The challenge before us is to create a robust ethics and policy framework for addressing them, and to do so quickly and in a way that is flexible enough to respond to technological advances. This might be achieved by an approach similar to the NIH's Recombinant Advisory Committee (RAC). The Nash case was a success story, but others have not been so successful. Other couples endured years of IVF and PGD in an attempt to create an immune-matched stem cell donor for their sick children, often spending their life's savings, but never had the Nashes' success. Many others will continue on similar paths in the hope of their own success—we need to create a path for success on the policy front, as well.

NOTES

1. A version of this chapter was published previously as "Creating a Stem Cell Donor: A Case Study in Reproductive Genetics," *Kennedy Institute of Ethics Journal* 14(1) March 2004; reprinted here with permission.

2. We are sensitive to concerns about patient and family confidentiality. The Nash family has been very public in their discussion of their daughter Molly's illness, giving many interviews for prominent print and television stories. They have explained their willingness to share their story in order to advance the medical and policy discussion of cases like theirs and to bring attention to needed research on Fanconi anemia.

3. Jeffrey Kahn is a current member of the board of directors of the NMDP.

4. Jeffrey Kahn is the current chair.

BIBLIOGRAPHY

Andrews, L.B., and N. Elster. "Regulating Reproductive Technologies." *Journal of Legal Medicine* 21, no. 1 (March 2000): 35–65.

Auerbach, A.D. "Umbilical Cord Transplants for Genetic Disease: Diagnostic and Ethical Issues in Fetal Studies." *Blood Cells* 20 (1994): 303–309.

Boyce, N. "A Law's Fetal Flaw." *U.S. News & World Report*, July 21, 2003, 48.

Belkin, L. "The Made-to-Order Savior." *New York Times Magazine*, July 1, 2001, 36.

Hoffman, D.I., G.L. Zellman, C.C. Fair, J. Mayer, J. Zeitz, W.E. Gibbons, and T.G. Turner Jr. "Cryopreserved Embryos in the United States and Their Availability for Research." *Fertility and Sterility* 79, no. 5 (May 2003): 1063–1069.

Knowles, L.P. "Reprogenetics: A Chance for Meaningful Regulation." *Hastings Center Report* 32, no. 3 (May–June 2002): 13.

Parfit, D. *Reasons and Persons*. Oxford: Clarendon, 1984.

Public Law No. 105-277, 112 Stat. 2681 (1998).

Robertson, J.A. *Children of Choice: Freedom and the New Reproductive Technologies*. Princeton, N.J.: Princeton University Press, 1994.

Robertson, J.A., J.P. Kahn, and J.E. Wagner. "Conception to Obtain Hematopoietic Stem Cells." *Hastings Center Report* 32, no. 3 (2002).

Wagner, J.E., S.M. Davies, and A.D. Auerbach. "Hematopoietic cell transplantation in the treatment of Fanconi anemia." In *Hematopoietic Cell Transplantation*, edited by E.D. Thomas, K.G. Blume, and S.J. Forman, 1204–1219. Malden, Mass.: Blackwell Science, 1999.

5

Optimizing Safety and Benefits of Genetic Testing: A Look at the Canadian Policy

Mylène Deschênes

The introduction of genetic testing into the Canadian market as another tool in the medical arsenal against disease is occurring at a fast pace. As more genetic tests are ready to leave the lab bench for the market, one may wonder about the type of control being exercised over these products to ensure that they are beneficial and safe for public use. The introduction of recent biotechnology innovations into the mainstream medical practice can either be left to the forces of a free market or be subject to different levels of scrutiny and control by the government or nongovernmental organizations. Canadians live in a society where risks are constantly assessed, evaluated, and, where possible, minimized. It is thus reasonable for Canadians to expect that the marketing of genetic tests will be under some form of surveillance. In fact, Canadians consider that biotechnology products require more surveillance than other marketable products and should meet higher scientific standards.[1]

Genetic testing challenges the current regulatory framework. At a period when the public is only becoming acquainted with the existence of genetic tests, specialists are still trying to fully appraise the social, ethical, and psychological implications that the diagnostic and, more importantly, predictive nature of these tests present. Given the possible consequences, we need to reexamine the existing normative framework to evaluate if it offers appropriate guidance. Critical questions need to be addressed. What is an adequate scheme of oversight for genetic tests? Who should be the lead authority? What standards are required to ensure a "safe" genetic test? We must also balance the need for surveillance and oversight against prompt access to technologies that may improve health. Undue hindrance of commercialization would not serve the public.

As one attempts to draft appropriate public-policy directives and provide guidance for the assessment and oversight of genetic tests, the choices made by decision makers must reflect society's perception of new technologies and how society manages risk. Such an exercise of drafting policy must therefore be based on objective and scientific measurement. The overall assessment of whether a particular technology is ready for consumer release and should be supported by a society can also take into account other elements, which reflect the values of a given society.

In this chapter, we first discuss public perception in Canada. Then, we delineate the current regulatory framework for the introduction of genetic testing into the Canadian market. Finally, we examine the approach adopted by the United States and compare it to the Canadian experience, evaluating suggestions made by various stakeholders to improve the current regulatory framework for genetic testing.

PUBLIC PERCEPTION OF BIOTECHNOLOGY AND GENETIC TESTING IN CANADA

The public's attitude and perception toward biotechnology and genetic testing can shape demand and impact public policy. In general, Canadians perceive advances in biotechnology favorably. In fact, in a recent survey conducted for the government, two-thirds of Canadians expressed support for technology.[2] They also believe that leading-edge technology is linked to societal progress, is inextricably linked to modernity, and is important to the country's future economic success.[3] In fact, biotechnology represents one of the fastest growing sectors of the Canadian economy. In 2001, Canadian biotechnology enterprises generated C$3.6 billion in revenues,[4] and it was estimated that in 2002 Canadian enterprises would generate C$5 billion.[5] Unsurprisingly, Canada fosters the development of scientific expertise and research infrastructures with initiatives such as the Canadian Foundation for Innovation, Genome Canada, and the Canadian Institutes of Health Research. Genetic experts are making important contributions to the improvement of health.[6]

Few surveys have thoroughly examined the Canadian perceptions of genetic testing. In Ontario, a survey conducted in 2001 showed that 89 percent of the participants were favorable to human genetic testing.[7] Overall, Ontarians believed that new genetic developments would mean healthier children free from genetic disease, although they expressed some concern about privacy-related issues.[8]

Fed by constant reports of groundbreaking discoveries that challenge the limits of knowledge, accelerated by the completion of the human genome project, the public has high expectations for genetic testing. Patients and interests groups increasingly demand prompt access to genetic testing.[9] However, in order to preserve the inherent trust of Canadians in this burgeoning field, introducing the fruits of biotechnology, such as genetic testing, must be approached with caution. Maintaining the public confidence is crucial to the successful integration of the products of biotechnology into society.

When questioned about biotechnology in general, most Canadians feel safe about the products accessible to them:

> Most Canadians believe that products on store shelves have been tested and are safe. Even though they have virtually no detailed understanding or knowledge of the federal Government's regulatory practices and imperatives, there is a general sense among Canadians that the systems are sound . . . . Most people want to know that government is trying to mitigate or reduce risks as society seeks to gain benefits.[10]

Canadians expect the federal government to assume responsibility for regulating genetic tests.[11] However, when interrogated on the current regulatory framework governing the field of biotechnology, only 28 percent of Canadians surveyed believed in the probable efficacy of governmental rules and regulations (compared to 40 percent in the United States).[12]

Canadians have mixed emotions about the introduction of new technologies. We are optimistic about biotechnology yet we express some mistrust in the current regulatory framework, which plays an active role in risk management. In this context, it is relevant to identify the current regulatory framework that enables some form of control and assessment over genetic testing, and to analyze whether it can, in fact, provide appropriate guidance in light of the multiple challenges raised by the introduction of genetic testing in daily medical practice. In the next section, we identify and analyze the Canadian regulatory framework for the premarket assessment of genetic tests and marketing strategies.

CONTROL OF GENETIC TESTING

In Canada, the federal government and the provincial or territorial governments can legislate on matters relating to health care.[13] Courts have ruled that any criminal matters related to falsification of health products is under the exclusive power of the federal government.[14] Provincial or territorial governments are responsible for the establishment, maintenance, and management of hospitals as well as anything related to property and civil rights.[15] Therefore, the analysis of the Canadian regulatory framework for genetic testing must take into account Canada's two levels of government and their respective powers over health care.

Another characteristic of the Canadian approach to health is the universal health-care system.[16] While the system has been under enormous pressures recently, health care is still offered in large part by the public sector. However, drugs and medical devices paid for by this universal coverage are usually subject to prior approval, that is, the service might be offered but paid for only if covered by provincial insurance.

Finally, an important legal and cultural element is the importance of personal autonomy and free choice. This has a tremendous impact on the doctor–patient relationship. Patients are relying on their health professionals for sound medical advice, but they do not expect an overly paternalistic attitude. Patients want to play a role in the decision-making process. In fact, people are generally encouraged to take more initiative and assume responsibility for their own well-being. Canadians demand rapid access to genetic tests and may even want to choose for themselves whether or not they deem a product ready for use.[17] This creates pressure on researchers, the industry, and any clearing agency looking into genetic testing.

Premarket Assessment

Genetic testing (with a health purpose) is offered in two ways.[18] First, the genetic test may be prepared as a "kit" and may thus constitute a medical device. Medical devices such as an in vitro diagnostic device are subject to federal scrutiny. Second, the genetic test may be prepared in-house by a laboratory ("home-brew") and offered as a laboratory service. In this case, it is subject to provincial authority over laboratories. In the following sections, we examine the regulatory framework for the premarket evaluation of both tests.

Genetic Tests as a Medical Device

Medical devices are subject to very specific evaluation and surveillance criteria by Health Canada under the Food and Drugs Act[19] and the Medical Device Regulation.[20] The evaluation system is based on the risk the diagnostic device presents: the greater the risk for the user and the public, the more demanding and stringent the regulatory requirements. Genetic tests have been classified as a level-three risk (a level-four risk represents the greatest risk).[21] Level-three in vitro diagnostic kits must comply with the general safety and effectiveness requirements set out in the Medical Device Regulation.[22] Furthermore, they must be licensed before being marketed.[23]

Health Canada reviews different aspects of a product before issuing a license. However, the evaluation of a medical device usually focuses on the physical properties of the device and the risk intrinsic to the device itself (for example, the risk that a pair of surgical scissors might break or that a catheter might perforate). In the case of an in vitro diagnostic device, particularly in the case of a genetic test, the risk evaluation reaches much further to include the evaluation of the consequences of getting an incorrect result, or simply of getting the result itself. The requirements for in vitro diagnostic devices are, in fact, more stringent. Health Canada requires "a summary of investigational testing conducted on the device using human subjects representative of the intended users and under conditions similar to the conditions of use."[24]

Health Canada's evaluation considers both the scientific and the clinical validity of genetic tests. Currently, Health Canada's guidelines cover basic scientific markers of quality and validity. Yet these standards may not be sufficient and properly tailored to the challenges raised by genetic testing. Perhaps we ought to consider broader criteria to evaluate whether a genetic test is effective, safe, and beneficial.

First, Health Canada falls short from assessing the clinical utility of a genetic test. In a text on in vitro AIDS test kits, the federal government indicated that it is not within its mandate to verify the extent of the use of the in vitro test kits:

> Health Canada's Medical Devices Bureau is responsible for ensuring that medical devices meet the "safety and effectiveness requirements" before licensing them for sale. Health Canada's assessment of safety and effectiveness is limited to assessing only the device's technical performance, although federal regulators do "require the manufacturers of point-of-care testing [kits] to submit data on consumer field evaluation to determine the device's performance when used by lay users, unassisted, following instructions provided in the labeling. The lay users should be representative of the target users for which the test is intended." However, in Health Canada's view, the Medical Devices Regulations "do not allow for the evaluation of these devices in terms of their impact [on] delivery [of test results] to the clients, their impact on current counseling methods, or psycho-social or other related issues."[25]

Clinical utility as well as the impact of a genetic test for a given user can have important consequences. For instance, predictive genetic tests, such as tests for Huntington's disease or breast cancer predisposition, provide life-changing medical information to the user. The risks–benefits evaluation leading to the decision of taking such genetic tests goes beyond a merely scientific analysis for the patient because the result can have social, economic, and psychological effects. Currently, this broad evaluation rests on the judgment of the medical professional.[26]

Second, there is no inquiry into the social or ethical value of a test. Society as a whole is not offered the chance to question the introduction of the product into the market. Instead, everything is left to the rules of the free market.

Health Canada's evaluation, although imperfect, offers a safety net of sorts that ensures that every genetic test kit is reviewed for inherent scientific and clinical validity. However, it is estimated that most genetic tests are currently being prepared as home-brewed. Consequently, they are not subject to Health Canada's scrutiny. The next section considers how those laboratory home-brewed genetic tests are evaluated.

Genetic Tests as a Laboratory Service

Tests done in a laboratory are not subject to evaluation by Health Canada. Each province regulates the laboratory services provided within its jurisdiction. For the purpose of this text, we examine only the situation in the province of Quebec. In Quebec, the regulation of laboratories offers no specific guidance for genetic tests. Laboratories must be licensed according to a given sector of activity, but genetic testing is not a recognized category.[27] The director of a laboratory is required to set up quality-control procedures and to comply with any demand made by the Quebec Public Health Laboratory. There are no regulatory standards and no specific control over genetic testing; neither is there formal premarket evaluation of such genetic tests before laboratories can offer them.[28] In the province of Quebec, the decision to start offering a home-brewed laboratory genetic test is not subject to an assessment by an external entity.

Control and Access

Genetic tests are generally conducted only under the supervision of a health-care professional. Genetic tests prepared as a kit cannot be advertised or sold directly to the public if they claim to cure, treat, or prevent a condition enumerated in an extensive list included in the legislation.[29] Genetic tests offered as a service by a licensed laboratory can only be provided under the prescription of a health-care professional.[30] Not only is a prescription required by law, but it is also necessary for the reimbursement of the procedure by the health-care system. Over-the-counter access is thereby constrained, although easy access through the Internet might lead to some important loopholes.

The mandatory inclusion of health-care professionals in this process provides a safeguard for the safe and appropriate use of such products or services. Amongst others, it ensures that consumers and patients know, before the test is administered, what consequences this result may have. It also ensures that the results are correctly interpreted. Finally, it prevents the proliferation of useless genetic tests and procedures.

This premise supposes that detailed and extensive information about genetic testing is available and that health professionals are up-to-date about genetic testing. Neither is always true.[31] In order to obtain a license in Canada, the safety and efficacy of genetic tests should be evidenced by prior clinical trials. However, such a requirement does not formally exist for laboratory products. For instance, we may know that a genetic test is scientifically valid, but we might have no idea about the prevalence of a disease in a given population. Thus, for the physician, it becomes difficult to evaluate the relevancy and appropriateness to suggest that an individual be tested.

The Canadian Situation

In short, the Canadian regulatory framework for providing oversight of genetic tests is in reality a patchwork of regulations applied to different types of genetic tests. It is not always properly tailored to assess genetic testing per se. Consumers do not expect to find different levels of reliability, efficiency, or usefulness depending on the nature of the test they choose (kits or laboratory services) or the province in which they require laboratory services. Moreover, the current framework does not provide for an assessment that considers elements other than scientific and clinical validity. Nevertheless, the most critical needs are in the area of laboratory services, where specific standards for genetic tests are currently absent.

REVISITING THE CURRENT POLICY

Given the gaps identified in the Canadian policy framework, we need to revisit Canada's current regulatory framework. The United States came to the same conclusion about five years ago, following an extensive report on genetic testing. Given the similarities between the Canadian and American regulatory framework, we first consider the suggestions made by the United States to modify its approach to genetic testing. We then discuss how Canada could improve its own regulatory framework.

Suggestions Made by the United States

Canada has just begun to analyze the regulatory framework governing genetic testing. In the United States, however, different initiatives have been under way since 1997 to study the concerns that have arisen due to the introduction of genetic tests into the market. In the United States the Food and Drug Administration (FDA) regulates in vitro kits with a legislative scheme that resembles that of Canada.[32] Laboratory services are regulated under the Clinical Laboratory Improvement Act (CLIA), mainly under the supervision of the Centers for Medicare and Medicaid Services (CMS) and the Center for Disease Control (CDC).[33] However, the legislation offers no genetic-specific regulations other than those targeting the specialty field of cytogenetics. A few states have undertaken to draft specific laboratory practice standards for DNA-based genetic testing. Such is the case with the State of New York, through the Clinical Laboratory Evaluation Program (CLEP).[34] This dichotomy in the regulatory approach poses problems similar to those that occur in Canada. The United States has undertaken, however, the study of these legislative gaps with the intention of proposing modifications.

A question that rapidly emerged from this undertaking was the determination of which governmental agency should take the lead on this issue. A few years ago, the FDA undertook controlling analyte-specific reagents used in home-brewed genetic tests. However, some authorities within the FDA indicated that it is not within their mandate or within the means of the FDA to control the oversight of all types of genetic testing. The overlap of jurisdictions will no doubt necessitate the involvement of different federal agencies including the FDA, CDC, and CMS.

In 1997, a newly assembled Task Force on Genetic Testing produced a report identifying some of the gaps in the regulations and recommending the creation of an advisory committee on questions related to genetic testing.[35] Then, in 1998, the Secretary's Advisory Committee on Genetic Testing (SACGT) was created. It issued recommendations related to enhancing the oversight of genetic testing.[36] SACGT stated that the "well-being of the population depends upon the rapid and broad availability of genetic tests" as well as their appropriate use. It added that:

> Based on the rapidly evolving nature of genetic tests, their anticipated widespread use, and extensive concerns expressed by the public about their potential for misuse or misinterpretation, additional oversight is warranted for all genetic tests.[37]

SACGT studied another important question: What criteria should be used to assess the benefits and risks of genetic tests?[38] SACGT recommended that analytical validity, clinical validity, clinical utility, and social consequences be the main criteria.[39] It recommended that the FDA be the federal agency taking the lead to review, approve, and label genetic tests, and that the Department of Health and Human Services (DHHS), under the CMS, should provide sufficient resources "to carry out expanded oversight of genetic tests." SACGT supported the creation of a specific category of specialty under the CLIA to ensure proper evaluation and assessment of genetic tests. Notice of intent was to be published in the Federal Register in 2002 but is still being discussed by the CDC and the CMS.

In the evaluation of risks and benefits from genetic tests, SACGT also expressed concern over the review of the social and ethical impact of genetic tests:

> Because FDA'S review will focus on assuring the analytical and clinical validity of a test, the agency's capacity to assess the ethical and social implications of a test may not be sufficient. The Secretary should consider the development of a mechanism to ensure the identification and appropriate review of tests that raise major social and ethical issues.[40]

Such an entity is not currently in place in the United States, but it might certainly offer a great opportunity for society to voice specific concerns regarding genetic tests before they become widely available on the market.

Improving Canada's Current Regulatory Framework

Canada has not created a task force similar to that of the United States. Although discussions are taking place, only a few documents analyze the current situation and make some recommendations. There is general recognition, however, that improvement is required.

Evaluating the Benefits and Risks of a Genetic Test

In Ontario, the question of how to evaluate the benefits and risks of a genetic test is being debated, and a recent report calls for the establishment of a collaborative network between the different provinces to develop optimum capacity in genetic testing assessment. According to this report, assessment should include "economic evidence, relative cost-benefit and medical efficacy studies conducted both pre and post approval."[41] Like the SACGT recommendations, it recognizes the need for an assessment that goes beyond a purely scientific evaluation. The notion of "medical efficacy" is far-reaching. It certainly comprises not only the criterion of scientific validity but also that of clinical validity and use. It also includes a whole new dimension in the risk-benefit analysis: the economic aspect of genetic testing. It is interesting to note that the United Kingdom made a similar remark in a report on laboratory genetic testing services. The suggestion went as far as making cost-benefit analysis part of the evaluation:

> Individual tests must be assessed in terms of analytical and clinical validity and clinical utility before introduced into the service arena. Quality assurance mechanisms and *cost-effectiveness should also be evaluated.* Mechanisms will need to be developed to enable this to happen. The ability to detect a variation in a gene does not in itself constitute an adequate reason for providing the test in service setting.[42] (emphasis added)

Such concern makes sense in countries with universal health-care coverage where risks and health-care costs are spread across the whole population.

Implementing New Standards

The evaluation of the safety and efficacy of genetic tests sold as kits is currently under the oversight of the federal regulatory scheme. There is an independent evaluation of the scientific and clinical validity of genetic tests. Similar evaluation should be done for home-brewed genetic tests. In this respect, specific genetic testing standards need to be adopted. This is urgent as the majority of genetic tests currently offered are home-brewed. From the consumer's point of view, there is absolutely no justification for such a discrepancy between home-brewed tests and kit tests.

As suggested by the SACGT, the evaluation of the inherent scientific validity of a given genetic test is insufficient to conclude that the balance of risks and benefits of the test is positive. We ought to consider other elements in the balance of risks and benefits such as the social consequences the results from a genetic test may have, for example, the discrimination or stigmatization a person may experience when it is known that the person has an increased chance of developing Alzheimer's in the future. Another element lies in the moral adequacy of a given genetic test, for example, a test that would look for a genetic predisposition to violence or criminality. These additional considerations have never been considered by Health Canada's device bureau in the evaluation of other in vitro diagnostic devices, and it is not its mandate to do so. Furthermore, the bureau does not have the means to complete such a task on all types of in vitro diagnostic devices. The same goes for laboratories. If such a global evaluation is required, we must consider other options for its implementation.

Offering a Global Evaluation for Genetic Tests

The Public and Physicians   The evaluation of the social and ethical consequences of using genetic tests can be left up to the judgment of the consumer or to the professional opinion of physicians and health professionals. In a society where people take a proactive role in the improvement of their health, citizens could decide for themselves about the use of a given genetic test. This is reflected in a recent public poll on biotechnology:

> And though Canadians expect ethical considerations to guide the development of these technologies, they are loath to allow the ethical standards of one person or group to determine whether a product should be allowed for all. *Once deemed safe, Canadians believe the individual consumer and the marketplace should be the sole determiners of the decision to purchase and use biotechnology products.* The preference of the vast majority is for individuals to make their own choices, based on their own ethical standards. The only situation where ethics trump other considerations, and where Canadians are prepared to accept a ban of an application on ethical grounds, is in the case of cloning human beings.[43] (emphasis added)

However, such judgments must be based on accurate and specialized information. It also presupposes that this information will be widely disseminated and easily accessible to the public. Faced with complex technology and test results that are difficult to interpret, this might not be the best option.

Physicians also play a key role in the introduction of this technology. They are in a better position than anyone else to evaluate the social and psychological impact of a genetic test for one of their patients. However, this is again conditional on the availability of sufficient quality information that allows them to base their professional evaluation on the most recent data about the use of genetic tests.

If commercialization is left to the free market, pre- and postmarket research and free public dissemination of the results should be mandatory. This will enhance the decision-making ability of physicians and the public.

The advantage of the universal health-care coverage Another option available to Canadians is to take advantage of the universal health-care system as another possible level of oversight. In Canada, the majority of health services are provided through the public system. This does not preclude a product from entering the private market. However, mass access to technology is clearly linked to its inclusion in provincial insurance plans. Over and above the simple clearance for marketing a given product or service in Canada on the basis of scientific validity, other considerations may be taken into account when it comes to evaluating its inclusion under provincial health-care insurance coverage. In fact, the incentive to be placed on the list for coverage could be used as leverage, for example, to persuade providers of genetic services to continue research on a given test.

In a country where health risks are shared, the scientific or clinical validity of a genetic test might not be sufficient to meet the required standards. One has to adopt a broad approach to risk-benefit assessment. For instance, not only is there a clear incentive to intervene in order to assess and control some of the risks, but there are also logical concerns for cost-effectiveness. It is therefore not surprising that concerns for cost-effectiveness were voiced both in Canada and in the United Kingdom, where a similar universal health-care system exists.

It has been suggested that we could take advantage of the medicare system to respond to social and ethical concerns related to the introduction of new technology through the process of listing medications, tests, and services for payment under the provincial health insurance plan.[44] This suggestion was made to address issues related to patent law. A similar approach could be adopted for genetic testing.

In its suggestion that the assessment of genetic tests take into account considerations other than purely scientific, SACGT fell short of identifying who would be making such an evaluation in the United States. In Quebec, an organization called Agence d'évaluation des technologies et des modes d'intervention en santé (l'AETMIS) is responsible, among others, for providing independent information about the introduction of new technologies into the medical mainstream. Such an agency could take the lead when it comes to making a global assessment of genetic tests.

CONCLUSION

Genetic tests are formidable technological innovations that will no doubt spur a quantum leap in the field of medicine. Striking a balance between safety and prompt access to technological innovations is an art! In the delicate task of finding an appropriate regulatory framework, different aspects need to be reviewed: finding an appropriate scheme to evaluate genetic tests, determining which considerations need to be taken into account in such evaluation, and, finally, establishing mechanisms to give voice to these concerns.

What constitutes a risk and when can we deem a product safe? Of course, scientific assessment is always the starting point, and at this level, some adjustments are required in Canada, particularly in laboratory testing. However, other considerations might also be relevant in a risk-benefit analysis. Genetic tests raise many social, ethical, and economic concerns, which can be relevant in a broad definition of risk.

In a country where health risks are commonly shared through a universal health-care system, it is only logical that some of these considerations be discussed prior to the endorsement of a test or product. Finding a way to incorporate these considerations into the process of approval of these products for public consumption goes beyond simple economic or scientific analysis. It would give not only individual consumers but society as a whole a say about some of the most difficult challenges raised by genetic testing.

NOTES

The author would like to acknowledge the help of Catherine Mosco, Madelaine Saginur, and Gabrielle Grégoire for the preparation of this chapter. The author also would like to express her deepest gratitude to the Stichting Porticus Foundation and the Genetics and Society Project.

1. For instance, this is the opinion collected in a survey conducted in Canada and in the United States about the public's perception on biotechnology. See Polara Research and Earnscliffe Research, *Public Opinion Research into Biotechnology Issues in the United States and Canada. Eight Wave Summary Report* (March 2003), 10, biotech.gc.ca/epic/internet/incbsscb.nsf/vwapj/Wave_8_Summary_Report.pdf/$FILE/Wave_8_Summary_Report.pdf.

2. Government of Canada, *Summary of Public Opinion into Biotechnology Issues in Canada* (Ottawa: Government of Canada, 2003), 3, biotech.gc.ca/epic/internet/incbs-scb.nsf/vwapj/Canada_and_Biotech_Research1.pdf/$FILE/Canada_and_Biotech_Research1.pdf.

3. Government of Canada, Public Opinion into Biotechnology Issues.

4. Statistics Canada, *Features of Canadian Biotechnology Innovative Firms: Results from the Biotechnology Use and Development Survey–2001*, www.statcan.ca/english/research/88F0006XIE/88F0006XIE2003005.pdf.

5. Government of Canada, *Follow the Leaders—Canadian Innovations in Biotechnologies*, (2002), www.innovationstrategy.gc.ca/cmb/innovation.nsf/vRTF/BioBook/$file/leader.pdf.

6. "Canadian researchers have made key contributions to genomics. Dr. Michael Smith won a Nobel Prize in Chemistry in 1993 for providing the world with one of the key tools for genomics research; Dr. Lap Chee Tsui discovered the gene responsible for cystic fibrosis; Dr. Jacques Simard of Laval University is a member of an international team that has discovered the permutations of genes that considerably increase the probability of breast cancer, particularly in young women; and Dr. Ron Worton identified the gene tied to muscular dystrophy." Reported by Dr. Henry G. Friesen, Chairman, Genome Canada, November 6, 2000, Toronto, www.genome-canada.ca.

7. Provincial Advisory Committee on New Predictive Genetic Technologies, *Genetic Services in Ontario: Mapping the Future* (Ontario: Government of Ontario, November 2001), 17.

8. Provincial Advisory Committee.

9. Timothy Caulfield, "Gene Testing in the Biotech Century: Are Physicians Ready?" *Canadian Medical Association Journal* 161, no. 9 (1999): 1122–1123.

10. Government of Canada, 2003, 5.

11. Government of Ontario, *Chartering New Territory in Healthcare, Genetics, Testing and Gene Patenting* (Ontario: Government of Ontario, January 2002), 56. Survey participants reported the following: 54 percent attributed the responsibility to the Federal government compared to 22 percent to the provincial government or 34 percent to an international Group.

12. Polara Research and Earnscliffe Research, 9.

13. *Constitutional Act 1867*, 30–31 Vict., R.-U., c. 31, sec. 91.

14. See R. v. Wetmore, (1983) 2. S.C.R. 284. See also Berryland Canning Co. Ltd. v. La Reine, (1974) 1 FC 91.

15. *Constitutional Act 1867*, 30 & 31 Vict., R.-U., c. 31, sec. 92 (7) et 92 (13).

16. *Canada Health Act*, R.S. 1985, c. C-6.

17. "Benkendrof and colleagues found that 95 percent of women in their study thought they should be able to get testing despite a physician's recommendation to the contrary. Similarly, a North American study found that 60 percent of those surveyed thought that they were 'entitled to any genetic service they can pay for out of pocket' and 69 percent thought that 'withholding any service was a denial of the patient's right,'" in Timothy Caulfield, 1122–1123.

18. Our focus is on genetic tests with a medical use diagnosis, treatment, or prevention of disease. It should be noted that paternity tests are not included under this category and are not discussed in this text.

19. *Food and Drugs Act*, R.S. (1985), c. F-27.

20. *Medical Devices Regulations*, SOR/98-282, *Canada Gazette* Part II, vol. 132, no. 11, SOR / 98-282 (May 7, 1998): 1645.

21. *Medical Devices Regulations*, sch. I, part. II, rule 4.

22. *Medical Devices Regulations*, sec. 10 to 20.

23. *Medical Devices Regulations*, sec. 26, 27 and 32. Licences are subject to yearly renewal.

24. *Medical Devices Regulations*, sec. 32 (3) h).

25. R. Elliott and R. Jürgens, *Rapid HIV Screening at the Point of Care: Legal and Ethical Issues* (Canadian Strategy on HIV/AIDS, 2000), 17.

26. The sale of medical devices directly to the public is forbidden if the device is presented as being able to cure, prevent, or treat a list of conditions presented in schedule 3 of the Food and Drug Act.

27. Regulation Respecting the Application of the Public Health Protection Act (now the *Act Respecting Medical Laboratories, Organ Tissues, Gamete, Embryo Conservation and the Disposition of Human Bodies*), R.R.Q., c. L-0.2, r.1, sec. 91, 93. However, laboratories must disclose the type of test they are doing in their license request.

28. Regulation Respecting the Application of the Public Health Protection Act. sec. 139–141.

29. *Food and Drugs Act* (Canada), Appendix A.

30. Regulation Respecting the Application of the Public Health Protection Act, sec. 136.

31. Timothy Caulfield.

32. *Food Drug and Cosmetic Act*, 21 U.S.C., c. 9, sec. 301 (As amended by the Food and Drug Modernization Act 1997).

33. CLIA, Title 42, ch. 6A, sub-c. II, part. F, sub-p. II.

34. NYCRR, title 10, sub-p. 58. New York State Department of Health, *Clinical Laboratory Evaluation Program: Guide to Program Requirements and Services* (Wadsworth Center), /www.wadsworth.org.

35. Task Force on Genetic Testing, *Promoting Safe and Effective Genetic Testing in the United States, Final Report* (Maryland, 1997), available on the website of the National Institute of Health (ELSI), www.nghri.nih.gov/ELSI/TFGT_final, published in Neil A. Holtzman and Micheal S. Watson, *Promoting Safe and Effective Genetic Testing in the United States, Final Report of the Task Force on Genetic Testing* (Baltimore: Johns Hopkins University Press, 1998).

36. Secretary's Advisory Committee on Genetic Testing, *Enhancing the Oversight of Genetic Tests: Recommendations of the SACGT, Final Report* (National Institutes of Health, 2000), www.od.nih.gov/oba/sacgt/reports/oversight_report.pdf. (Hereinafter: SACGT report).

37. SACGT report, ix.

38. SACGT report, vi.

39. SACGT report, viii.

40. SACGT report, 31.

41. Government of Ontario, 85.

42. National Health Services, Human Genetics Commission, *Laboratory Services for Genetics* (August 2000), www.hgc.gov.uk, 55.

43. Government of Canada, 2003, 3.

44. Richard Gold, "Biomedical Patents and Ethics," *McGill Law Journal* 45 (2000): 432. The author makes this suggestion in the context of patent law.

6

Experimentation on Human Embryos: The Bioethical Discussion in Europe with Special Attention to Austria and Germany

Heinrich Ganthaler

Research on human embryos promises to bring great benefits in that it might lead to the development of effective therapies for a wide range of human illnesses that are currently difficult or impossible to treat, especially nervous system diseases, cancer, immune diseases, and diabetes. Legislation concerning research on human embryos is nevertheless very restrictive in many European countries. In the Convention on Human Rights and Biomedicine of 1997, Article 18, the European Council states two conditions on research on human embryos:

1. Where the law allows research on embryos in vitro, it shall ensure adequate protection of the embryo.

2. The creation of human embryos for research purposes is prohibited.[1]

An additional protocol to the convention approved in 1998 prohibits the reproductive cloning of human beings.[2]

In June 2002 the European Parliament adopted the Sixth Framework Programme for Research and Technological Development for the period 2002–2006.[3] As the European Commission states, research programmes must conform to the current legislation in the countries where the research is carried out. No funding is allowed for research into human cloning for reproductive purposes or research intended to modify the genetic heritage of human beings. Concerning stem cell research, the European Parliament (in conformance with a decision of the European Council) decided to give priority to research on adult stem cells and not to fund the production of embryos solely for research purposes including by means of somatic nuclear cell transfer (that is, therapeutic cloning). However, a compromise was adopted concerning research on already existing human embryos and stem cell lines. The ban on certain kinds of research in some countries should not prevent the European Union (EU) as a whole from supporting such research in countries where it is allowed.[4] In opposition to this European Parliament Regulation the Council of Europe also proposed not to fund research on supernumerary embryos, spare embryos leftover from infertility treatment.

At the national level, research on human embryos is currently strictly prohibited in Ireland, Norway, Poland, Germany, Austria, and Italy.[5] In France, the law that prohibits embryo research is under review. The use of supernumerary embryos for research purposes is permitted in Finland and Spain. The United Kingdom and Sweden provide a very liberal legislation according to which not only the use of supernumerary embryos but also the creation of embryos for research purposes (for example, by therapeutic cloning) is permitted. In the Netherlands and Belgium, a bill is being prepared that prohibits (with many exceptions) the production of embryos for research purposes but permits research on supernumerary embryos. In general the condition imposed on research on human embryos is the prohibition of research after the fourteenth day of the existence of the embryo.

This chapter concentrates on the debate in Austria and Germany. First, I summarize the opinions of the Austrian and German National Committees on Bioethics concerning research on human embryos. Second, I discuss some arguments for and against research on human embryos from a philosophical point of view.

THE DEBATE IN AUSTRIA AND GERMANY

In Austria the production of embryonic stem cell lines is currently prohibited by law. The import of stem cell lines, however, is not expressly prohibited. Therefore, according to Austrian legislation, EU funding of related research is also not prohibited. In an opinion published in May 2002, the Austrian Bioethics Commission at the Federal Chancellery states that in general all kinds of efforts to develop new therapies for human diseases that are currently untreatable deserve to be promoted, provided they are acceptable from an ethical and legal point of view.[6] Concerning human embryonic stem cell research, the commission states that its ethical legitimacy depends on the moral status that is attributed to the human embryo, that is, whether or not the human embryo is regarded as having a right to life and human dignity. Another important question is whether there is an essential difference between embryos created by oocyte–sperm fertilization and those created by cloning, that is, by the transfer of the nucleus of a donor cell into the denucleated oocyte.

Regarding these issues, the members of the commission did not reach consensus. Nonetheless, they agreed on the following matters: (1) the commission welcomed the Council of the European Union decision that places priority on research on adult stem cells over research on embryonic stem cells; (2) the commission expressed support for the European Parliament decision that excludes reproductive cloning, the production of embryos for research purposes, and the modification of the genetic heritage of human beings from funding by the European Union; (3) the commission also expressed support for the decision of the Council of the European Union to exclude research on supernumerary embryos as well as research in the field of somatic cell nuclear transfer (therapeutic cloning) from funding by the European Union.[7]

On the other hand, the members of the Commission disagreed on whether or not research on already existing embryonic stem cells should be prohibited. However, the majority of the members spoke out in favor of research on already existing embryonic stem cells, given the following provisions:

1. It should involve high-level, peer-reviewed research projects without any other alternatives. For basic research, animal embryonic stem cell lines or human adult stem cell lines are to be used wherever possible.
2. Only those stem cell lines should be used which are derived from embryos which have exclusively been produced for medically assisted reproduction (IVF), but which can no longer be implanted.
3. The nonpurchased and informed consent of the donors is necessary.
4. Until further notice, only those stem cell lines that already existed before a given date may be used, so that the creation and destruction of supernumerary embryos created through in vitro fertilization (IVF) treatment are not encouraged for purposes of stem cell research.
5. Research projects shall be assessed by an independent interdisciplinary commission that cooperates with the local ethics commissions.
6. All results, including negative research results, should be reported and must be published, similar to the publication of adverse events.[8]

The opposing minority emphasized instead the need to protect all forms of human life and to respect the dignity of all human beings, thereby raising, among other concerns, the following objections against legalizing research on already existing supernumerary embryos:

1. If the existing stem cell lines available do not prove to be sufficient for research, there is the risk that there will be an increased demand for the deliberate creation of embryos for research purposes.
2. The restriction of research to preexisting stem cell lines that were obtained before a certain time period cannot be substantiated objectively nor justified ethically.
3. The high level of a research project and its therapeutic goals cannot justify the destruction of human embryos.

4. To date, research has neither demonstrated academic proof that embryonic stem cells are irreplaceable for the development of future types of medical treatment, nor has it specified any real clinical benefits that go beyond just a mere general increase in knowledge.
5. The patentability of human embryonic stem cell lines, as well as the resulting economic exploitability, would pave the way for research directed at embryonic, as opposed to adult, stem cells.
6. Ethically required publication as well as the assurance of sound perception of research results are at present difficult to enforce.
7. A recommendation for research with embryonic stem cell lines would exert pressure for the repeal of the current prohibition on embryonic research in Austria.[9]

Like the Austrian Bioethics Commission at the Federal Chancellery, the German National Ethics Council was also not able to reach consensus on matters relating to the use of human embryos for research purposes. The existing German law on embryo protection prohibits the fertilization of an ovum for purposes other than its reimplantation in the donor, but, similar to Austrian legislation, there is no rule prohibiting the use of human embryos or stem cell lines imported from other countries.

For this reason the German National Ethics Council discussed the question as to whether or not the import of human embryos or stem cell lines to Germany should be prohibited. The majority of the members spoke out in favor of the permissibility of the import of human embryos and stem cell lines, while a minority strongly opposed it. In an opinion published in 2001, the council lists a number of arguments for and against research on human embryos, thereby showing up some apparent inconsistencies in German law.[10] One such inconsistency relates to the fact that in Germany there exists a rather liberal abortion law, according to which an abortion, given certain provisions, may be procured up to three months past conception. This means that according to the existing German law an embryo in vitro is currently better protected than an embryo or fetus in vivo. According to the proponents of research on human embryos it is impossible to justify such an unequal treatment of embryos in vivo and embryos in vitro. Moreover, they argue that the argument brought forward by the opposing minority, according to which a human embryo has a right to life from the moment of conception, is not convincing. Although the right to life is a fundamental right of the German Basic Law (constitution), according to its usual interpretation it does not apply to human beings before birth. For this reason, preimplantation genetic diagnosis (PGD) is also ethically acceptable, as the German National Ethics Council states in an opinion published in January 2003.[11]

PHILOSOPHICAL CONSIDERATIONS

Summarizing the debate on embryo research in the Austrian Bioethics Commission at the Federal Chancellery and the German National Ethics Council we find that a minority in both committees defends a doctrine, sometimes called the "Doctrine of the Sanctity of Life." According to this doctrine—mainly defended by the Roman Catholic Church—human life is sacred, and human beings have an inalienable right to life from the moment of conception on. This doctrine has far-reaching consequences. According to this doctrine not only abortion and the use of abortifacients (nidation inhibitors that prevent fertilized ova from being implanted) but also the destruction of human embryos for the purpose of research are strictly forbidden. Thereby it does not matter whether the embryo in question is a supernumerary embryo leftover from infertility treatment, an embryo created in vitro by oocyte-sperm fertilization, or an embryo created by cloning for research purposes (therapeutic cloning). Moreover, in vitro fertilization in itself should be prohibited in so far as it leads to the creation of supernumerary embryos that are to be destroyed after some time. For the same reason, PGD aimed at selecting human embryos (that is, implanting only healthy embryos and discarding those with a defective gene) is forbidden.

The opinion of the majority on both committees differs from the opinion of the minority only in so far as it accepts research on supernumerary embryos or, in the case of Germany, the import of human embryos or human stem cell lines. At the same time it strongly rejects the creation of human embryos only for research purposes. In this context two questions arise: Is such a rigorous view as defended by the proponents of the "Doctrine of the Sanctity of Life" justifiable? Is it ethically justifiable to differentiate between supernumerary embryos leftover from infertility treatment and embryos created for research purposes?

To start with the first question, we may ask, What in general makes it wrong to take the life of a human being? The most obvious reason why people want to have their lives protected is the fact that people value their lives and are interested in their lives for various reasons. As ethicist John Harris puts it:

> Each of us will have our own reasons for valuing our own lives and each of us is able to appreciate that the same is true of others; that they too value their own lives. What we have in common is our *capacity* to value our own lives and those of others, however different our *reasons* for doing so may be or may seem to be. I believe those rather simple, even formal features of what it takes to be a person—that persons are beings capable of valuing their own lives—can tell us a good deal about what it is to treat someone as a person. They can tell us how to recognize other beings as people, and they also tell us why it's wrong to kill

such creatures against their will. They are people because they are capable of valuing life, and it's wrong to kill them because they do value life.[12]

Moreover, human beings can experience pleasure and pain, have various desires, are capable of self-awareness and self-control, have a sense of the future, and of the past and have relationships with other people—all such attributes being indicators for what we may call "personhood." In other words, to take the life of a human being means to kill a person and to deprive it of its future, to thwart all its goals and unfulfilled desires. But obviously, an early embryo (for example, an eight-cell embryo used for research) is not yet a being with the mental qualities just mentioned. Devoid of a nervous system or a brain it can neither experience pain or pleasure nor have any interests or desires about its future. It does not even know of its existence and therefore cannot have any interest in staying alive. For this reason research on human embryos seems to be acceptable. I call this argument *the argument from personhood*.[13] The usual objection to this argument is that a human embryo, although it is not yet a person, has the *potential* to become a person and that it is wrong to do anything that prevents it from fulfilling this potential. For this reason any kind of destruction of human embryos (in the course of research as well as in the course of PGD) is strictly forbidden. I call this argument *the argument from potentiality*. But why should potentiality count?

First, it may be objected that a potential person does not have the same rights as an actual person has. That something will become X is not sufficient reason for treating it as if it were in fact X. Although the potentiality of the human embryo may justify the attribution of a special status to it, it does not justify the ascription of the legal status of a human being after birth to it.[14]

Second, it may be objected that ova and sperm before conception are already potential persons. However, in this case, who would argue that we should not prevent them from fulfilling their "potential," that is, that there is a duty to reproduce? In other words, there is no intrinsic difference between preventing the development of a potential person by contraception and preventing the development of a potential person by destroying an early embryo or preventing its implantation.[15] From this point of view there is no reason to prohibit research on human embryos, and for the same reason the use of PGD aimed at the selection and destruction of human embryos with a defective gene cannot be regarded as ethically wrong.

However, in the case of PGD there is another argument to be considered. The selection and destruction of embryos with a defective gene in the course of PGD, it is often said, is, like abortion, to be regarded as an act of discrimination against disabled people. But it is one thing to infringe upon the right to life of an actual person (for example, by killing her against her will) and quite another thing to prevent the development of a potential person (which does not yet have an interest in its own life). Presupposing that early embryos do not yet have a right to life, to prohibit the development of an early embryo with a defective gene by its destruction is no more discriminating than prohibiting the development of a possibly disabled human being by contraception.[16]

On the other hand, this does not mean that potentiality does not count at all. Provided we foster the development of a potential person to the point where it develops sentience and actual interests, we have a duty to take into account these future interests and to respond to them in an adequate way. In this sense we have a duty to care for future persons and to respect their right to life, whether they are disabled or not.

But what about the creation of human embryos specifically for research purposes? Is there an ethically relevant difference between research on supernumerary embryos and embryos created specifically for the purpose of research? According to a famous principle of Kantian ethics, we never may use a human being solely as a means to an end.[17] To create an embryo specifically for research purposes, it is often argued, is to instrumentalize a human being, whereas in the case of a supernumerary embryo, the embryo in question has not initially been seen as a means to another end. For this reason the creation of an embryo solely for research purposes is ethically unacceptable. I call this argument *the argument from instrumentalization*.

But does it really make sense to apply Kant's principle to a totally non-sentient, nonrational, and nonautonomous being that can have no ends of its own? At least from the point of view of an interest-based ethics, Kant's principle cannot be applied to a being without any interests of its own. Therefore the argument from instrumentalization is also not convincing.[18]

Nevertheless, there are good reasons to subject the creation of human embryos for research purposes to certain basic constraints. Technically speaking, there are two ways of creating human embryos in vitro: (1) creating a human embryo by oocyte–sperm fertilization or (2) creating an embryo through the transfer of the nucleus of a donor cell into a denucleated oocyte (i.e., therapeutic cloning). The main advantage of using stem cells derived from embryos created by therapeutic cloning would be that such stem cells are autologous to the recipient and therefore not subject to immune rejection.

In both cases consent of the donors of biological material should be required and it should be prohibited that embryo donation be a commercial transaction. Moreover, therapeutic cloning must be prevented from becoming merely a first step to reproductive cloning. The main reason for this is that attempts at the reproductive cloning of human beings would entail serious risks. From several attempts to clone animals (like the famed cloned sheep Dolly) we know that such attempts are successful only in a small percentage of cases and that cloned animals usually suffer from various diseases, while at the same time have a shortened life expectancy. Therefore, it would be irresponsible to carry out such experiments on human beings and a general ban on reproductive cloning is to be welcomed.[19]

But this is not the case with therapeutic cloning, which is not intended to implant embryos created in this way into a woman's uterus and to foster their development to a stage where they exhibit consciousness and are capable of valuing their lives. In other words, whereas attempts at reproductive cloning bear a high risk in harming future persons, the use of stem cells derived from embryos created by therapeutic cloning does not harm any future person. There are, of course, other ethical complications raised by therapeutic cloning. The technique of nuclear transfer is still very uncertain. In order to improve this technique, a large number of eggs will have to be expended. What women would donate eggs for this purpose, and are there sufficient safeguards against commercialization?[20]

To conclude, taking seriously the argument from personhood, there are no good reasons to accept the "Doctrine of the Sanctity of Life" and to prohibit research on human embryos. With respect to the ethical status of early human embryos there are also no good reasons to make a difference between embryos leftover from infertility treatment and embryos specifically created for research purposes. Neither the argument from potentiality nor the argument from intrumentalization is genuinely convincing. For the same reasons the selection of embryos resulting from PGD is not ethically wrong. Regardless of the ethical status of human embryos, there may, of course, be other reasons to prohibit the creation of human embryos solely for research purposes, such as to prevent the exploitation of women as egg donors.

Because of their potential to develop into persons, human embryos deserve to be treated respectfully and must not be destroyed arbitrarily. But there is also a duty to relieve human beings from suffering and to help sick people. Weighing the possible benefits to children and adult human beings (actual persons) from embryo research against the destruction of human embryos at an early stage of development (the destruction of potential persons), research on human embryos, provided there is no scientific alternative, as well as the selection of human embryos resulting from PGD are ethically justified.

NOTES

1. Council of Europe, *Convention for the Protection of Human Rights and Dignity of the Human Being with Regard to the Application of Biology and Medicine: Convention on Human Rights and Biomedicine*, European Treaty Series Number 164 (1997).

2. Council of Europe, *Additional Protocol to the Convention for the Protection of Human Rights and Dignity of the Human Being with regard to the Application of Biology and Medicine, on the Prohibition of Cloning Human Beings*, European Treaty Series Number 168 (1998).

3. European Commission, *The Sixth EU Framework Programme for Research and Technological Development in Brief* (2002).

4. European Commission, 7.

5. An overview is given in UNESCO: International Bioethics Committee (IBC), *The Use of Embryonic Stem Cells in Therapeutic Research. Report of the IBC in the Ethical Aspects of Human Embryonic Stem Cell Research* (Paris 2001), 4–5.

6. Bioethics Commission at the Federal Chancellery, *Decision of 3 April and 8 May 2002. Opinion of the Bioethics Commission on the Issue of Stem Cell Research in the Context of the EU's Sixth Framework Programme for Research, Technological Development and Demonstration Activities as a Contribution towards the Realization of the European Research Area* (2002–2006).

7. Bioethics Commission at the Federal Chancellery, 2.

8. Bioethics Commission at the Federal Chancellery, 3–4.

9. Bioethics Commission at the Federal Chancellery, 5.

10. German National Ethics Council, *Opinion on the Import of Human Embryonic Stem Cells* (December 2001).

11. German National Ethics Council, *Genetic Diagnosis Before and During Pregnancy* (January 2003).

12. John Harris, *The Value of Life: An Introduction to Medical Ethics* (London: Routledge, 1991), 16–17.

13. For a similar position see H. Kuhse and P. Singer, *Individuals, Humans, Persons: Questions of Life and Death* (Germany: Sankt Augustin, 1994), and N. Hoerster, *Ethik des Embryonenschutzes* (Stuttgart 2002).

14. Harris, 1991, 11.

15. Harris, 1991, 11–12, and H. Kukse and P. Singer, "Abortion and Contraception: The Moral Significance of Fertilization," in H. Kuhse and P. Singer, 83–106.

16. See John Harris, "Should We Attempt to Eradicate Disability?" in E. Morscher, O. Neumaier, and P. Simons, eds., *Applied Ethics in a Troubled World* (Boston: Dordrecht, 1998), 105–114.

17. I. Kant, *Grundlegung zur Metaphysik der Sitten* (Riga 1785, 1786), abgedruckt in W. Weischedel, ed., *Immanuel Kant's Werke in zehn Bänden*, Band 6 (Darmstadt, 1968), 66f.

18. H. Kuhse and P. Singer, "Beginning of Life: The Issue of Moral Status," in Kuhse and Singer, 67–82.

19. For this and other reasons the Austrian Bioethics Commission at the Federal Chancellery also opposed reproductive cloning. See the opinion published in Bioethics Commission at the Federal Chancellery, *Decision of the Bioethics Commission at the Federal Chancellery of 12 February 2003. Interim Report on So-called Reproductive Cloning with Regard to a Detailed Opinion on the Application of Human Cloning, Embryo Protection and Embryo Research, Pre-implantation Diagnostics as well as Additional Issues Concerning Reproductive Medicine.*

20. For this argument see Swedish National Council on Medical Ethics, *Statement of Opinion on Embryonic Stem Cell Research* (Stockholm 2002), paragraph 7.2.

7

The Cultural Challenge of Biotechnology in Post-Communist Europe

Larissa P. Zhiganova and Yuri M. Gariev

This volume focuses on two main topics: biotechnology and culture. This chapter looks at these two subjects as applied to a group of countries that we call post-Communist Europe, which does justice to both the geography and cultural space that they share. Culture is a transcendent realm where shared interests, values, and aspirations interact with each other in a continuum in which the past, present, and future are intricately interwoven.

There is little question that biology and its industrial applications—biotechnology—have entered the cultural landscape for good. This is an increasingly accepted assumption that has been instilled by nearly two centuries of concentrated interest in biology and the life sciences. Several factors have served to develop this interest. The theory of evolution made humans look at ourselves as biological actors. We have reached a growing understanding of the material underpinnings of heredity all the way from Gregor Mendel's experiments with peas to Watson and Crick's double-helix model of DNA, the carrier of genetic information, and onward to contemporary genetic research epitomized by the Human Genome Project. The understanding of the inner workings of microorganisms and living cells has enabled us to intervene directly in these processes and mend what we assume to be wrong. Finally, modern computer technology has made it possible to analyze massive volumes of complex but quantifiable biological data. All this knowledge has progressively made its way from desks and labs to hospitals, factories, farms, pharmacies, and shopping lists. It is not surprising that the growth of biotechnology has also penetrated the political arena.

Much of the biological research of the twentieth century was pioneered in the United States, a country endowed with scientific talent and powerful sources of finance. Now, the United States postures itself as the international leader in biotechnology, setting standards for the rest of the world and im-posing its products on other markets. Since 1995, large U.S. biotech compa-nies have been inundating farmers with a steady supply of corn, soybean, cotton, and vegetable seeds genetically modified to resist pests and weeds or to harbor enhanced nutritional values. The conquest of large chunks of North American markets through sophisticated marketing techniques was followed by aggressive policies to promote the products and technologies overseas.

Another major industrial application of biotechnology where the United States remains unsurpassed is biomedicine. Here, the focus is currently on "regenerative" medicine that aims to reverse the destruction of tissues result-ing in debilitating conditions of which Parkinson's disease, Huntington's chorea, and multiple sclerosis are only a few examples. Unlike traditional forms of medicine, regenerative medicine depends on human cells and tis-sues for its raw material. Gene therapy, biotech pharmaceuticals, and cloning techniques are other major biomedical fields where the United States is the leader. One of the sources of the growing demand for regenerative and genetic therapy has been an increase in the incidence of neurodegenerative disorders and other conditions that some researchers tend to link with the deteriorating "genetic load" of the human race.

CULTURAL DIMENSIONS OF BIOTECHNOLOGY

Biotechnology is a definitive reality of the globalizing world. Any economic reality that achieves critical mass in a given society is bound to spill into the domains of ethical, religious, aesthetic, social, legal, and political sensibili-ties, that is, is likely to gain a cultural dimension. Although "biotechnology" is an umbrella term for thousands of different technologies, their culturally sensitive component comes down to a limited number of issues that can be broken down into two basic categories: physical and metaphysical issues. The physical issues have to do with a variety of safety concerns, including the risks of biotech interference in the food chain, the unknown effects of genetic modification of agricultural plants and animals, and the conse-quences of treating animals and humans with genetically engineered sub-stances. Other risks stem from novel biomedical interventions (like cell or gene therapies) and assisted reproductive technologies (ARTs), as well as techniques that may compromise environmental sustainability of animal, plant, and even human populations. The "metaphysical" issues include de-rivatives of culturally conditioned notions of the freedom to interfere into life and nature, especially the embryo and fetus—the raw material of stem cell, fetal, and other types of therapy—or to enhance desired physical features in humans through genetic modification or cloning. Other issues are making headlines as well, especially the legal protection of intellectual property rights with regard to live objects or parts thereof such as genes or cells. Biotechnology challenges the notions of distributive justice, that is, the availability, ownership, and use of its fruits as well as breeds fears of genetic discrimination.

Any technology may be viewed as a tool to advance the values and interests already present in a given society. Therefore, a discussion of the impact of biotechnology on the fabric of culture and individual lives in a particular region should yield a more general, critical look at these values and interests.

THE CULTURAL CONTEXT OF POST-COMMUNIST EUROPE: PAST

Looking at the political map of Eastern Europe, one sees the sprawling landmass of the Russian Federation, the former constituent republics of the Soviet Union (Ukraine, Belarus, Moldova, Lithuania, Latvia, and Estonia) and a number of countries that were under Soviet political, economic, and cultural domination for a significant part of the twentieth century: Poland, the Czech Republic, Slovakia, Hungary, Romania, Albania, and former republics within the federation of Yugoslavia.

Some background generalizations can be made about this part of Europe. Partly because most of these territories were for long periods the political backyards of larger entities, the mostly agricultural economy was based on a feudal mode of managing land, property, and workforce. Therefore, the region has performed poorly, in the eyes of the West, in terms of building up institutions of private ownership, market economy, representative and accountable government, and a civic society in general. Because the region has been dominated by large entities—Hapsburg, Russian, Prussian, and Ottoman empires, later by the Soviet Union, and now by the European Union (EU) and the Russian Federation—none of these small countries, with the exception of Russia, have a tradition of viable sovereignty. The predominant faiths in the region have been Orthodox Christianity, Roman Catholicism, and Islam. The absence of Protestantism impeded secularization and the ensuing boom of capitalism.

What sets Russia apart from the rest of post-Communist Europe are its vast territory, its recent imperial past, and its unique and assertive civilization. Russia now spreads over a territory of 6.6 million square miles in both Europe and Asia. Until 1917 the Russian Empire included parts of modern Poland, Finland, and the Baltic countries. Since the inception of the Russian state in the ninth century, it was growing in territory and governed autocratically. A prominent role in pre-Soviet Russia belonged to Orthodox Christianity. Ancient-rite liturgical worship, diligent prayer, iconography, and theological perspectives on church, individual, society, and the state were an integral part of the people's day-to-day existence. Russia's self-confidence started eroding when Tsar Peter the Great started his West-oriented reform in the early eighteenth century. Since then, Russian political and philosophical thought has been roughly split into "Westernizers" and "Slavophiles."

The Bolshevik revolution of 1917 was a turning point that exploded the Russian state, church, and society. The Soviet Union, with its intellectual foundation in a rather unorthodox interpretation of Marxism, enforced a cohesive system of materialist answers to all life's questions. The Soviet system eventually subdued most of Eastern Europe.

Several aspects in Marxist-Leninist ideology and practice are relevant to the object of this study. The USSR practiced an extreme form of statism, seeking to regulate the economy, political life, education, and private lives. This disproportionate role of the state, with some modifications, was also practiced in the socialist countries of Eastern Europe.

The Soviet state produced a most dramatic spiritual and intellectual fissure in Russia, echoed in the satellite countries, which was the abrupt scrapping of the history of Christianity accompanied by inculcation of "scientific atheism" as the ultimate truth. This was a cohesive system of ideas, which, according to one practitioner of Soviet propaganda, "put in the place of faith a Communist conviction, based on a scientific world-view and Marxist-Leninist doctrine, i.e. [a conviction in the form of] knowledge of the theory and practice of scientific Communism which becomes part of the inner world of the person and is manifested in his everyday activities."[1]

Religious nihilism fulfilled some important political functions. One of them was the reorganization of reproductive behavior and family life. A curious aspect in Socialist policy was the state endorsement of abortion upon request. Abortion had been outlawed in Russia for centuries until 1920 when the Soviet health and justice ministries issued an unprecedented decree permitting "the free-of-charge procedure of artificial termination of pregnancy within the Soviet hospitals where the utmost harmlessness of the procedure [was] ensured."[2] The Bolsheviks used abortion as a policy tool. Through liberating the woman, the government tried to subvert the Christian idea of family that had previously been the foundation of the much-hated Russian statehood. In this context, the murder of the entire Romanov royal family in 1918 rather than one monarch was truly symbolic. The rushed industrialization of a previously agricultural country called for a buildup of the workforce, necessitating female labor. Abortion was outlawed in 1936 to overcome a demographic decline and was reinstated in 1955 under Nikita Khrushchev when it became clear that the ban had given rise to a spread of unauthorized abortions that were taking an increasing toll on women's lives and fertility.

Also, the first decades of Soviet history featured a number of bans on entire fields of research, including genetics, that had been a field of vibrant endeavor for Russian scientists since the early twentieth century. Their boundless country with its unparalleled biological diversity provided an excellent venue and resource for experimentation. After 1917, their motivation ensued from the need to modernize the agricultural sector in naturally adverse conditions and to improve healthcare. Nikolai Vavilov, a plant geneticist, breeder, geographer, taxonomist, and physiologist, funneled the findings of his endless expeditions, field experiments, and lab research into the enhancement of Soviet plant agriculture. Grigoriy Levitskiy established the "classical" school of chromosome morphology research, and his manual on cell genetics was the first one in the world. Nikolai Koltsov's "matrix principle" of the genesis of chromosomal molecules proved prophetic. Alexander Serebrovskiy and Georgiy Nadson advanced the study of genetic mutations with the use of radiation. The synthesis of genetics and evolutionism was the focus for Sergey Tchetverikov. Boris Astaurov explored the agricultural applications of hybridization and asexual reproduction. Yuriy Filipchenko and Sergey Davidenkov promoted medical genetics and genetic counseling along with an agenda of population-improvement measures. Russian researchers pioneered many genetic concepts, including such fundamental ones as "gene pool."[3] In this light, Stalin's repressive policies were a painful blow for Russia's biology for decades to come.

The excessively powerful, ideologically driven Soviet state outlawed private free enterprise, suppressed spiritual and political freedoms, muted the civil society, routinized abortion, and put roadblocks in the way of some of the most promising fields of scientific research.

THE CULTURAL CONTEXT OF
POST-COMMUNIST EUROPE: PRESENT

In the late 1980s, the Socialist International went west, almost literally, when Communist ideology proved economically and politically bankrupt and the Eastern European nations started looking at their Western neighbors for role models and integration options. The EU now has a membership of fifteen states with ten others slated for accession within the next few years and still more aspiring to the same. The North Atlantic Treaty Organization (NATO) was developed as a trans-Atlantic defense alliance. The small Eastern European countries now have the opportunity of economic and political integration into the EU and NATO. For the former Soviet republics, the main integration option is the Commonwealth of Independent States, a loose bloc of nations dominated by the Russian Federation.

The post-Communist Eastern Europe is now challenged by a complex range of issues relating to biotechnology, which has already made deep inroads into European diets, healthcare, industry, and the environment. However, the point of departure for post-Communist Europe is still largely determined by its special historic baggage.

First and foremost, the ban on genetic research in the first decades of Communism dealt a heavy blow to the current material base of the Eastern European biotech sector. Russia is a case study of the gap between biotech brainpower and its economic realization. The engine of life sciences research and development in Russia is represented by a network of "institutes" and "centers" under the administrative auspices of the Russian Academy of Sciences and the health and agriculture ministries. They now have to adjust to economic uncertainty and a bare minimum of state funding. At the same time, there is a growing demand for biotech products that are seen by some actors in the Russian agribusiness sector as an engine of productivity growth. To meet this demand, 95 percent of the biotech products used by meat producers and 75 percent of those used in plant agriculture are imported from the West.[4] It comes as no surprise that biotechnology has the air of foreignness, which complicates a healthy public interest in and analysis of biotechnology as a whole.

Second, the long-time absence of the free market, private ownership, and entrepreneurial culture complicates the commercialization of biotech innovations. The marketing of know-how requires complex skills that scientists are only beginning to pick up. According to Abercade Consulting, a Russian consultancy specializing in innovation marketing, the transformation of promising technology into a product is impeded by the insufficient understanding of all questions relating to business modeling, such as realistic marketing, investment analysis, cultivation of functional partnerships, and, finally, the business idea itself.[5]

Third, the inadequate development of civil society is also a factor that prevents any serious debate on biotech-related issues in these countries. General public awareness of assisted reproductive technologies (ARTs), gene therapy, stem cell research, genetic modification of plants and animals, or the Human Genome Project remains low. A mid-1990 survey of professional and lay attitudes to ARTs in Russia lamented the "inadequate public awareness of this issue, weak coverage of positive and negative aspects of ARTs in domestic popular literature, lacking discussion of alternative ways to solve problems with childbirth other than the ARTs."[6] Professional opinions were hardly more comforting. Similar conclusions are echoed in many opinion polls. A survey aiming to identify public attitudes to "the moral acceptability of using novel genetic technologies to improve human heredity" found that Russians generally have greater trust in "technological methods of solving human problems . . . a sort of 'technological enthusiasm' that has been undermined in the West in the past thirty years by sharp criticism."[7] Articles and books on biotechnology, bioethics, and biopolitics, though not unusual, are still somewhat rare in the mainstream press, although the quality of what does appear is generally good.

Fourth, the widespread acceptance of abortion in the Soviet Union is still deeply felt in the Eastern European societies. In Russia, the apparatus of pregnancy termination in the post-Soviet health-care system has all the makings of an efficient "industry," with flourishing associated fields of research such as fetal-tissue therapy. On the sidelines of the health-care system, a shadow abortion industry has prospered with impunity. For example, in 1989, 1.5 million unauthorized abortions were performed in Russia.[8] A major outcome has been the desensitization of the society to the problem of the status of the embryo. It doesn't take a crystal ball to predict that such a society is likely to be less ethically discerning about stem cell and fetal tissue therapies.

Fifth, Eastern Europeans had to go through a traumatic disruption of historic continuity twice in the twentieth century. The Communist revolution and its consequences uprooted the Eastern European societies, and the collapse of the Socialist system left people free to shop for their ideas and values in the marketplace. The bottom line so far has been the lack of a value system that could form the groundwork of legislation. For example, Russian legislation and legislative discourse on assisted reproductive technologies, abortion, sterilization, genetic engineering, or biomedical experimentation displays a thorough understanding of technical complexity. Yet this discourse is not accompanied by an understanding of ethical implications. In the North Atlantic world, respect for ethically sensitive issues has been reinforced by ubiquitous review boards. However, by way of example, morally sterile Russian regulations, a leading bioethicist complains, do not "contain a provision on mandatory ethical review of any research project involving research on humans."[9] Nevertheless, although mature discourse in Eastern Europe is found wanting, we should do justice to the rising enthusiastic voices that seek to expose the relevance of biotechnology. These include academics, some nongovernmental organizations (NGOs), and churches.

There are two major groups of scientists in Eastern Europe who contribute to the promotion of biotech knowledge in Eastern Europe. One group includes scientists who have a stake in pushing their innovations into the market. They are learning to market their discoveries and inventions; to attract financing for their research, development, and marketing efforts; and to protect their know-how through intellectual property laws. The scale of their discourse includes trade conferences, other publicity and outreach events, lobbying, and awareness-raising through the Internet. The challenge concerning the protection of intellectual property rights has been a priority. Russia has its own federal patenting system and is a member of the Eurasian Patent Convention. The federal system is currently in a state of reform and revision: new rules slated to be adopted shortly are expected to elaborate the evaluation of novel biotech "inventions" such as cells, plants, and animals as well as to streamline the requirements for patenting more familiar objects of scientific study, like nucleic acids and proteins. It is also expected that biotech inventions will be given their own separate niche in the regulatory framework. The rules of the Eurasian Patent Office already refer to "biotechnological products" as inventions that can be patented.

The second group focuses on the subtle cultural derivatives of biology. In Russia, these scientists promote bioethics, biopolitics, and biosociology, seeking to prove the relevance of their concepts to Russian policy makers. Biopolitics was first popularized in Russia by Alexander Zoob, and it later became the subject of interest of a few other researchers. Moscow State University's biology department has a biopolitics and biosociology section that pursues close ties to the Biopolitical International Organization, the Gruter Institute, the European Sociobiology Society, and other institutions worldwide. It is yet too early to speak of any socially or culturally relevant contribution to large-scale public debate and policy. Bioethics as an academic theme, however, is making inroads into mainstream education through the publication of quality course materials, textbooks, and anthologies. A chapter on biotech-related material is now a usual part of any general ethics textbook.[10] A more public-oriented discourse is found in the activities of such NGOs as Greenpeace. Greenpeace Russia is now active in protesting against the unchecked contamination of the Russian food market with products containing genetically modified ingredients. Greenpeace informs the public about the incongruities between the wording and enforcement of Russian legislation. For example, the Ministry of Health has issued a directive that requires the labeling of any product containing "genetically modified ingredients" of "5 percent or more," which, Greenpeace in Russia asserts is not enforced at all (no. 30, 2003). Greenpeace enjoys a reasonably positive public reputation in the Russian Federation, and most of its actions are covered in the mainstream news media with a measure of sympathy, but its campaign against genetically modified food is still in the incipient phase.

Academics and NGOs draw mainly on Western experiences and expertise. More "domestic" players in this area are the traditional churches. The Russian Orthodox Church, for example, holds a holistic and protectionist view of family, society, state, and individual spiritual practice and discipline based on a pious, ascetic interpretation of scripture and tradition. In Western terms, the Orthodox Church is strictly pro-life, an ardent opponent of abortion and euthanasia. A recurrent topic for both pen and pulpit, the treatment of abortion relies heavily on notable church figures from the near and distant past and contemporary scientific perspectives. There are also several "single-issue" pro-life groups, such as Zhizn, Vita, and Pro-Life.

The church has also been vocal in addressing genetic diagnostics and therapy, assisted reproduction, and cloning. There are some ideas and areas of concern that are shared by many Orthodox authors. One such area is the physical and moral stress placed upon women by abortion, ARTs, sterilization, and the generally competitive social environment. The Church criticizes the concept of family planning as a demographic attack on the decreasing Russian population. Seeing in it a way to minimize the number of children, encourage the use of contraception, and promote Western-style sex education for the youth, Orthodox scientists, clergy, and laypersons preach a return to the family values distilled in the Orthodox Church for centuries. The Orthodox perspective also finds distasteful the reductionist focus on "sperms" and "oocytes" rather than on the "married couple."[11]

Bioethics received priority treatment in the Foundations of the Social Concept of the Russian Orthodox Church adopted by the National Council of Bishops in August 2000.[12] The council reiterated the unacceptability of abortion (except in cases where the mother's survival is at stake), contraception, cloning, and fetal therapy and expressed a very skeptical view of assisted reproductive technologies and the notions of reproductive rights and family planning. With regard to genetic diagnostics, the church welcomed the new possibilities of early detection of disease but warned against the use of acquired information as a justification for abortion.

Gene therapy was recognized as a positive achievement with great potential to prevent and cure some of the most debilitating diseases. However, the bishops made a crucial point here that is an important part of Orthodox thinking about human nature. The Orthodox mind held particular views on heredity long before genetics began to explain the material aspect of this process. Virtue and sin are not restricted in their impact to the one person in whom they originate. Rather, an individual wrong victimizes the environment of the person and may pass on to his or her offspring—sometimes in the form of hereditary disease. Alcohol, nicotine, and drugs are obvious examples, but the reality is that most actions that the Orthodox Church considers sinful are, quite objectively, destructive for the body and mind. "Pride–vanity–frustration–irritation–psychosomatic disorder" is quite a logical sequence. The church draws attention to the moral aspect of heredity and shuns a materialist philosophy of healing that ignores the original spiritual causes of disease.

These perspectives along with their proponents and many others in the society will be further enriched or challenged by the inevitable inflow of products, technologies, and services into post-Communist nations. It remains to be seen if these imports will be paralleled by a growth in awareness of and opposition to ensuing perceived problems.

THE CULTURAL CONTEXT OF POST-COMMUNIST EUROPE: FUTURE CHALLENGES

For the natives of this exciting region, the main challenge will be to overcome the subtle consequences of Communist heritage: the chronic lagging behind in research, the underdeveloped market and civil society, the disregard for human life exemplified by the unquestioning acceptance of abortion, and a lack of a strong and shared value system. The prime responsibility for confronting these deficiencies rests with the Eastern Europeans themselves, but in the process of globalization this responsibility is to be shared with the other members of the international community.

The substance of a responsible approach will be the reinforcement of a checks-and-balances system in which developments in industry and science can be checked by the civic institutions and the state to enforce a genuine, practical respect for a diversity of perspectives. The civil society and the state should learn to understand that they are potentially either the first beneficiaries or the first victims of any scientific or technological advances. Such a realization will only come through promoting a culture of free information and open debate, letting every voice speak out and argue. This process should result in the crystallization of a comprehensive value system or systems that can, in their turn, inform and guide the progress of research, development, and legislation in the field of biotechnology.

On the way to this kind of balanced framework, a series of distortions are taking place. We can see, for example, the deliberate marketing of questionable products to Eastern European consumers who are unaware of their accompanying concerns, such as genetically modified soybeans or nutriceuticals, or the export of fetal tissue from Russia where the public is largely but not completely indifferent to the subject. Distortions of this kind occur because the international and domestic players are not always guided by even a basic code of conduct and may be tempted to capitalize on the deficit of local knowledge or interest.

Two things are needed to avoid these distortions. First, products and services should only be introduced into new markets in conjunction with open discourse related to novel biotechnical elements. Second, it is imperative to engage in a continuous exchange of ideas with local voices. These agendas bring us to the point where the freedom of our choice, moral and otherwise, comes into play. The attempt to honor local interests and values is a *voluntary* leap outside the parochial, corporate, and orthodox into the realm of opposing ethical, religious, economic, and social frameworks that transcend the immediate moment, that exist in the past, present, and future at the same time, and are the flesh and bones of what we call "culture."

NOTES

1. Nikolai Krasnikov, *Pravoslavnaya etika: proshloe i nastoyashcheye* (Moscow: Politizdat, 1981), 85.

2. A.Y. Ivanyushkin, V.N. Ignatyev, R.V. Korotkikh, et al., *Vvedeniye v bioetiku: uchebnoye posobiye* (Moscow: Progress-Traditsiya, 1998), 202.

3. D.K. Belyayev and V.I. Ivanov, eds., *Vydayushchiyesya sovetskiye genetiki* (Moscow: Nauka, 1980), 60.

4. O. Pak, "Problemy investirovaniya v biotekhnologicheskiye razrabotki dlya selskogo hozyaystva" (2003), at www.rusbio.biz/ru/pak02.shtml.

5. Biochimmash, Biotekhnologiya i biznes, III Mezhdunarodnaya konferentsiya, 21-22 aprelya 2003 goda (Abercade Consulting, 2003), 8.

6. Boris Yudin, ed., *Bioetika: printsipy, pravila, problemy* (Moscow: URSS, 1998), 394.

7. Yudin, 416.

8. Yudin, 162.

9. Ivanyushkin et al., 378.

10. See, for example, T.V. Mishatkina and Y.S. Yaskevich, *Etika* (Minsk: Novoye Znaniye, 2002), 304–341.

11. M.B. Ovchinnikova, *Tekhnika zhizni, kotoraya vedyot k smerti* (Moscow: Favor, 2002), 133.

12. Council of Bishops of the Russian Orthodox Church, *Tserkov' i mir. Osnovy sotsialnoy kontseptsii Russkoy Pravoslavnoy Tserkvi* (Moscow: Danilovskiy blagovestnik, 2002), 137–155.

8

Why Is This Gene Different from All Other Genes? The Jewish Approach to Biotechnology

Edward Reichman

The opinions expressed herein represent those of the Orthodox Jewish tradition. While the voice of Orthodox Judaism is not monolithic and, indeed, a plurality of approaches, within accepted boundaries, is the norm, one can nevertheless distill immutable principles and values deriving from the Bible, Talmud, and legal codes, which inform the discussion and guide the decisions of rabbinic authorities. Debate and nuanced textual interpretation are the hallmarks of Jewish legal discourse, and this chapter provides a glimpse, albeit limited, into these processes. This chapter is intended solely for educational purposes and not for legal extrapolation. Each individual case in Jewish law should be presented to a rabbinic authority for adjudication.

JEWISH LAW AND APPROACHES TO MODERN BIOTECHNOLOGY

For Jewish law, the starting point is the Bible, or Torah, which was given directly from God to Moses and the Jewish people at Mount Sinai. But the word "Torah" has a broader meaning, referring to Jewish teachings and traditions that have been passed down throughout the centuries. In the second century of the Common Era, Rabbi Judah the Prince compiled the Mishna, a collection of rabbinic teachings transmitted since the giving of the Torah. In the fifth century, the Talmud, an expansion of the Mishna, containing thousands of pages of rabbinic discourse, was completed. This work, along with its myriad of supercommentaries, still serves as a focal point of rabbinic legal discussions and is studied daily throughout the world in the halls of the Yeshiva (school for Torah study). The method of Talmudic analysis is unique and involves partners discussing or arguing over passages of text for hours at a time. Indeed, it is common for only a few pages of Talmud to be completed in a semester's course.

In the twelfth century, Maimonides codified the law. The codification process was further developed in the sixteenth century, culminating with the authorship of the definitive code of Jewish law, known as the Shulchan Aruch. Rabbinic literature subsequent to this period has largely taken the form of questions and answers on specific practical issues, and this corpus has come to be known as the responsa literature. In approaching issues of contemporary law, and in particular Jewish medical ethics, rabbinic authorities draw upon all of the aforementioned sources.

In the story of creation, God enjoins man to "fill the land and conquer it" (Genesis 1:28). Jewish theological predisposition is not only to welcome but to aggressively pursue new technologies that improve our lives and our world. Yet, we cautiously evaluate each technology on its own legal merit to determine if its exploration or application entails violations of Torah principles. The Jewish approach is guided by the legal analysis of any issues stemming from the technology and its application. If, after careful and thorough analysis, no clear legal impediment exists to the use of the technology, then its use may be permitted. However, overarching philosophical or theological concerns, buttressed by rabbinic texts or the weight of traditional interpretation, may militate against the use of certain technologies, despite their technical, legal permissibility.

FUNDAMENTAL LEGAL PRINCIPLES

There are a number of legal principles that serve as the foundation for decisions in the field of biotechnology.

The License to Heal

Medieval commentators debated the propriety of the practice of medicine in the face of a divine decree of illness, and clearly resolved that the practice of medicine and the treatment and cure of human disease is a mandate of humanity. The Talmudic discussion about the "license" to heal has also defined the parameters of permitted medical intervention. Curative or palliative interventions clearly fall under the rubric of the medical license. Whether body-enhancing procedures are included in this category is a matter of debate. Some authorities forbid purely cosmetic procedures as their attendant risk cannot be justified as therapeutic. Others, however, allow certain cosmetic procedures that are considered psychologically therapeutic. This distinction may impact the approach to biotechnology since many interventions straddle the fence between therapeutics and enhancement.

The Sanctity of Life

Jewish law considers life to be of infinite value and zealously guards the sanctity of all human life, from birth to death, whether conscious or comatose, whether of sound mind or profoundly demented. The preservation of human life is paramount in Jewish law, and all biblical and rabbinic prohibitions (except murder, illicit sexual relations, and idolatry) are suspended to facilitate its preservation.

The Preservation of Human Life—*Pikuach Nefesh*

While saving a human life (*pikuach nefesh*) is clearly permitted and even obligatory, the license to violate Torah law for the potential or possibility of saving a life was addressed in the eighteenth century by Rabbi Yechezkel Landau. The question was posed by a family of a man in London who had succumbed to postoperative complications of bladder stone surgery. The surgeons requested an autopsy in hopes of possibly benefiting future patients presenting with the same condition. Jewish law, however, forbids the desecration of the corpse and requires immediate burial of the complete body. Rabbi Landau was asked if these laws could be suspended for the possibility of saving a life in the future. In addressing this request for a postmortem examination, Rabbi Landau argued that the laws relating to the treatment of the corpse (immediate burial, prohibition of desecration and deriving benefit from a corpse) may only be suspended for a *choleh lifaneinu*, "an ill patient before us," but not for the theoretical possibility of future benefit. The benefit must be direct and immediate. This decision of Rabbi Landau has been widely applied in the area of Jewish medical ethics and is the focal point for discussions on when Torah law can be suspended for the possible preservation of life (*pikuach nefesh*). Limitations of its application have been debated. For example, whether one must have in mind a specific patient who will potentially benefit, or whether potential benefit for unspecified individuals is sufficient to meet the threshold, is a matter of rabbinic debate. In addition, given our advanced communication and rapid dissemination of information, some have attempted to broaden the definition of "an ill patient before us." It has been argued that, since information gleaned from an autopsy may save lives and can be immediately posted on the Internet, perhaps autopsies, or other prohibited procedures, should generally be allowed. This, however, is not the accepted interpretation.

Scientific research does not meet this threshold of *choleh lifaneinu*. Although research may ultimately lead to the cure of human disease, the benefit is neither immediate nor direct. Indeed, the majority of scientific research yields no clinical benefit. As such, one may not violate Torah law in order to perform scientific research. For example, if research on embryonic stem cells would violate Torah law, one could *not* argue that it could be permitted for the sake of *pikuach nefesh* on the grounds of saving lives in the future as a result. To be sure, most scientific research entails no specific Torah violations whatsoever. Furthermore, according to Jewish law, the fetus in utero is not granted the status of a full human life, and its exact legal status during different embryological or developmental stages is a matter of rabbinic debate. This fact not only impacts on the legality of abortion, but also on the status of the pre-embryo and the permissibility of violating Torah law in order to preserve its existence.

LEGAL APPLICATION

Following are examples of the application of the aforementioned principles, in conjunction with the entire corpus of Jewish law, to some areas of biotechnology. This list is by no means exhaustive and is intended to provide a flavor of the approach of Jewish law to current issues in the use of biotechnology.

Abortion

While I do not wish to revisit the abortion issue here, it is important to reiterate the Jewish position on abortion inasmuch as one of the consequences of evolving prenatal genetic testing relates to the decision whether to abort the affected child. While it is true that the fetus in utero does not enjoy the full status of a human being, according to Jewish law it is nonetheless forbidden to wantonly perform abortions. The nature of the prohibition of abortion continues to be a matter of rabbinic debate, and depending on the various legal positions, abortions may be permitted in limited, extenuating circumstances.

Jewish law explicitly allows abortion for maternal indications (i.e., to save the life of the mother). This permissibility does not extend, however, to cases of abortion for subsequent use of the fetal tissue, organs, or stem cells to save another individual. There is rabbinic debate regarding abortion for fetal indications (i.e., the case of an impaired fetus). Even those in the minority who allow abortion for fetal indications generally restrict their rulings to fetal diseases that are fatal and untreatable, such as Tay-Sachs disease. If Tay-Sachs disease should one day be treatable by enzyme replacement or genetic therapy, then the legal basis for abortion in this case would no longer apply.

In the discussion on abortion, the relevance of the gestational age of the fetus has received some attention. The Talmud describes a stage of fetal development prior to forty days gestation during which the fetus has a different legal status. For example, during the times of the Temple in Jerusalem, when a woman spontaneously miscarried, she was subject to the laws of ritual purity and impurity just as if she had given birth to a full-term infant. The one exception to this law was the pre-forty-day fetus. If a woman miscarried prior to forty days of gestation, no ritual impurity was generated, since the fetus at this developmental stage was considered "mere water." It is debated whether this pronouncement about the pre-forty-day fetus has relevance to the modern abortion debate. While some authorities maintain that the early developmental stage mitigates the severity of the prohibition of abortion, others see no difference between an abortion before or after forty days of gestation.

This debate may impact on the harvesting of stem cells from pre-embryos, which occurs at a stage prior to forty days of embryological development. However, the forty-day debate may be restricted solely to the fetus in utero, which is considered a "potential" life. In any case, the continued expansion of prenatal genetic testing will undoubtedly test the limits of these rulings. Whether one should allow abortions for a child bearing a gene for an adult-onset disease or a gene for disease predisposition has not received sanction from any rabbinic authority and awaits further rabbinic analysis. Preimplantation genetic diagnosis (PGD) may circumvent some of the legal issues of abortion. Nevertheless, this still has its own attendant legal issues, which will be discussed in the following section.

Assisted Reproduction

The fact that the obligation to procreate, "be fruitful and multiply," is the very first *mitzvah* (commandment) in the Torah reflects its significance in the Jewish tradition. While one is not required to use assisted reproductive procedures to fulfill this obligation, these technologies may be used on a case-by-case basis provided they do not entail Torah prohibitions. One legal obstacle that pervades all discussions of assisted reproduction is the procurement of the male reproductive seed. There is a prohibition against the wasteful emission of male reproductive seed (i.e., outside the context of natural intercourse), and this concern guides many of the rabbinic discussions on this issue. The majority opinion allows sperm procurement to treat infertility, since the ultimate objective is to produce a child and to thereby fulfill the commandment to "be fruitful and multiply." This is not deemed a wasteful emission. However, sperm procurement for the production of pre-embryos from which stem cells would be harvested for research is not allowed.

Most authorities allow any interventions, such as artificial insemination (AI) or in vitro fertilization (IVF), using the gametes of a husband and wife and consider the resulting progeny to be the couple's legal child. When introducing donor gametes, however, either sperm or egg, authorities disagree as to the permissibility. Issues such as legal adultery and bastardy, the definition of maternity in cases of a gestational host (i.e., a woman who serves only as the gestational mother, while the egg was donated from another woman), and psychological manifestations weigh heavily in rabbinic discussions. IVF procedures invariably yield surplus embryos. This fact is not a legal impediment to the procedure, and authorities have discussed whether the surplus embryos may be discarded.

Genetic Screening and Genetic Testing

In addition to the license to heal, which applies to health-care practitioners, the patient has a reciprocal Torah obligation to seek medical care and practice preventative medicine. This principle is particularly applicable in the Jewish community to premarital genetic carrier testing. One of the first genetic diseases identified with the Jewish community was Tay-Sachs disease. Rabbis initially debated the advantages of testing, and Rabbi Moshe Feinstein, a great Talmudic sage of his generation, considered testing a moral obligation. Not being tested, he maintained, was like closing your eyes to obvious danger. One organization, Dor Yesharim, currently performs routine premarital testing for diseases such as Tay-Sachs, Canavan's, Bloom's, and cystic fibrosis. Through such testing, the incidence of Tay-Sachs disease in the Jewish community has dropped precipitously.

With the preliminary completion of the Human Genome Project, the application of genetic screening will surely continue to expand. Issues for consideration include the following: For which diseases should one screen? Which populations should be tested? And to whom do the results belong and to whom should they be divulged? Jewish law has addressed these issues, but the challenges will continue to grow with the broader application of testing.

Testing for genetic predisposition is perhaps the most vexing problem with respect to genetic testing since many questions still remain unresolved. A small percentage of Jewish women carry the BRCA1 or BRCA2 genes, which have been identified with a predisposition to breast and ovarian cancers. What percentage of these women will develop the disease and at what age, and how effective are prophylactic measures? All these questions are the subject of intense scrutiny and will affect how Jewish law approaches the issue.

One is surely obligated to take measures to prevent disease. However, if the disease cannot be prevented, one may not be obligated to obtain information about predisposition to that disease. Current medical literature indicates that prophylactic measures can prevent cancer in BRCA-positive women, but these measures vary from medications to imaging studies to radical surgeries. While one may incur a measure of risk to achieve a cure of an existing disease, what risk and how much sacrifice must one take in order to possibly prevent the onset of a disease? Prophylactic medicines and imaging studies may have only a small risk. Surgical procedures are clearly in a different category. On the one hand, if the surgery could prevent disease, it would be permitted, and perhaps even obligatory in Jewish law. On the other hand, if the chance of acquiring the disease is small and the risk of the procedure is high, it might be forbidden, according to Jewish law, to undergo the procedure. While other factors are involved, these decisions depend heavily on current scientific knowledge and will surely work together with the scientific advances.

Genetic Therapy

As mentioned previously, there is a firmly established mandate that humanity has the license to heal. This surely includes measures that would involve genetic manipulation to cure existing diseases, such as severe combined immunodeficiency syndrome. In principle, Jewish law does not differentiate between somatic cell therapy (that is, therapy that affects only the treated individual) and germ-line therapy (that is, therapy involving reproductive cells, the results of which could be transmitted to one's progeny). The fact that the intervention may affect subsequent generations does not, in itself, preclude its use. However, it must be established that the intervention is not only therapeutic for the treated individual, but also not detrimental to future progeny. Due to the current lack of sufficient scientific data that addresses genetic expression in future generations, any pronouncement on this matter is premature.

Status of the Human Embryo and Stem Cell Research

As discussed previously, Jewish law zealously guards the sanctity of human life—all human life—and Torah laws are suspended to preserve it. However, this legal protection is only afforded to those who meet the definition of "human life" according to Jewish law. Does a human embryo in the laboratory meet the definition of "human life"? If so, then one would not be allowed to use this embryo for research, as it would be tantamount to homicide. Homicide is forbidden, even if it means saving another life. As the Talmud states, "who is to say your blood is redder than his!" Is an embryo's "blood" indeed as red as a born, living human being? At what stage does human life begin, according to Jewish law, such that one can suspend other Torah prohibitions to preserve it?

To illustrate, imagine the following case. A fertility bank reports a malfunction of the embryo preservation system on the Sabbath. Failure to act promptly will invariably lead to the destruction of hundreds of human embryos. Yet repair of the problem requires the violation of the Sabbath. May one violate the Sabbath to preserve the embryos? The answer to this question will shed light on the legal status of the embryo. We will answer it by reviewing the rabbinic

approach to other cases of Sabbath violation, beginning with the adult and going in reverse chronological order. To save the life of a born human being, one
may, and indeed is obligated to, violate all the Torah precepts including keep-
ing the Sabbath (excluding those precepts barring homicide, illicit sexual relations, and idolatry). The operative legal principle in Jewish law is termed
pikuach nefesh. *Pikuach nefesh*, saving or preserving a human life, as mentioned previously, is a principle of paramount importance in Jewish law and
pervades many areas of Jewish medical ethics. According to Maimonides, if a
practical case presents itself, the sage of the community should specifically be
the one to violate the Sabbath in order to demonstrate the importance of human life in the eyes of the law.

Does this legal protection apply to a fetus in utero? The Mishna teaches
that if a woman's life is in danger during labor, as long as the fetus is still in
utero, one may dismember the fetus limb by limb to save the mother. However, the Mishna adds, if the head of the fetus exits the birth canal, one cannot harm the fetus even to preserve the mother's life. The reason: One cannot extinguish one life to save another. The implication from this teaching is
that while the fetus remains in utero, it is not accorded the legal status of a
full human life. If such is the case, may one violate the sacrosanct Sabbath to
preserve a fetus in utero? The current accepted opinion is that one may violate the Sabbath to preserve the life of a fetus, even if the mother's life is not
at risk. The decision is based, in part, on the notion that while the fetus is not
an "actual" life, it is a "potential" life, and we may violate one Sabbath in order for the fetus to ultimately observe future Sabbaths after birth.

What about the pre-embryo? Is this definition of potential life applied to
the pre-embryo such that one may violate the Sabbath to preserve the existence of the pre-embryo? The answer here is a clear and emphatic no. All authorities agree that the Sabbath may not be violated to preserve an embryo
outside the woman's womb, as the so-called pre-embryo, in the eyes of the
law, does not have a status of even potential life. In the absence of human
intervention (i.e., implantation), the pre-embryo will not develop into a full-
fledged human being. Does this mean that all human embryos can be discarded or used for research with impunity? Not necessarily. Since the pre-
embryo ultimately derives from the reproductive seed, and, as mentioned
above, there is a prohibition against the wasting of this seed, some authorities forbid its wanton destruction.

Given the lack of human status of the preimplantation embryo, many rabbinic authorities allow for its use in stem cell research. Some voice the concern that using the pre-embryo for stem cell research, which requires destruction of the embryo, is a violation of the prohibition of wasting the male
reproductive seed. Others counter that the aforementioned prohibition applies only to wasteful emission of the male reproductive seed, not to already
emitted seed, irrespective of its use in the fertilization of an embryo.

All those who allow use of pre-embryos for stem cell research limit this sanction to *existing* pre-embryos only, and they do not permit the *prospective* creation of embryos for stem cell research. It is forbidden to emit sperm for the purpose of creating an embryo that will be destroyed for research purposes. As stipulated above, sperm procurement is approved for assisted reproduction in order to fulfill the commandment to "be fruitful and multiply," but one may not violate Torah prohibitions (e.g., against wasteful, nonprocreative emission of male reproductive seed) for research purposes. However, once such an embryo is created, there is no additional prohibition in destroying it. Some authorities place additional limits on which surplus embryos can be used for research, restricting research to embryos either not viable (i.e., unable to fully develop if implanted) or not designated for implantation.

Preimplantation Genetic Diagnosis (PGD) and Genetic Selection

Rabbinic authorities embrace the use of PGD for the treatment of disease because it falls within the purview of the obligation to heal. The use of PGD to produce a "savior sibling" who would serve as an umbilical-cord-blood stem cell donor to treat the disease of an existing sibling would also fall within the parameters of *pikuach nefesh*, as it is a lifesaving procedure. Furthermore, a child is produced accomplishing the fulfillment of the commandment of "be fruitful and multiply." However, Jewish law does not advocate the indiscriminate use of PGD for trait selection or enhancement or for sex selection (excluding sex selection for sex linked diseases).

To some extent, preimplantation genetic diagnosis circumvents the legal issues relating to abortion. While abortion is considered legally objectionable, with PGD no fetus is destroyed and the tested embryos are not destroyed. While it is true that embryos are created that may never be implanted, failure to implant these extra embryos, which do have the status of human life, is not legally objectionable, especially since the procedure was performed to produce a disease-free child. Jewish authorities do not consider this practice to either disregard or devalue the sanctity of life. As discussed previously, the pre-embryo does not possess the status of human life, and one who destroys it would not be culpable of infanticide. Some authorities would recommend against the active destruction of the pre-embryo, while they would consider passively allowing its decomposition to be less objectionable.

Does this legal advantage of PGD over prenatal testing and abortion allow the use of PGD to select out for fetuses carrying genes for adult-onset diseases or for predisposition to diseases, conditions for which abortion would generally be prohibited according to Jewish law? Despite some legal advantages, PGD still requires in vitro fertilization and the procurement of male reproductive seed. This procedure is not free of legal concern and is only allowed for a legally justifiable benefit. Selecting out for potential or future disease of progeny may fulfill this criterion, but this matter awaits further rabbinic analysis.

Cloning

The process or "mechanical," aspects of human cloning present no major legal obstacles from a Jewish perspective. Neither somatic cell manipulation nor the subsequent implantation into a woman's uterus constitutes a violation of Torah law. Furthermore, since the reproductive seed is not needed in the process, the concern about the prohibition of wasting male reproductive seed, an issue for assisted reproduction, is irrelevant here. However, the low efficacy and potential adverse outcomes of human cloning are legal concerns that would lead Jewish authorities to reject any human cloning at this time. Even should these medical concerns be resolved, other legal and theological concerns may supervene. Prospectively creating people of legally ambiguous lineage and who may suffer profound social and psychological complications may preclude any future acceptance of cloning despite perfection of the procedure from a medical perspective. Some rabbinic authorities have found a limited use for human cloning, such as in the treatment of infertility.

CONCLUSION

Judaism is a rich, text-based tradition that addresses all aspects of human existence. Contemporary rabbinic authorities have applied its teachings to the fascinating and increasingly complex field of biotechnology. Despite the antiquity of Jewish law, its principles are equally applicable today as they were thousands of years ago. I am inclined to agree with the words of Benjamin Franklin:

> The Progress of human knowledge will be rapid and discoveries made of which we have at present no conception. I begin to be almost sorry I was born so soon, since I cannot have the happiness of knowing what will be known a hundred years hence.[1]

Whatever discoveries will be made in the coming centuries, the rabbis will continue to address them from a Jewish perspective.

NOTE

1. Benjamin Franklin, "Letter to Sir Joseph Banks," July 27, 1783. Full text at www.tpromo.com/gk/files2/benj-ban.htm.

BIBLIOGRAPHY

Baskin, J. "Prenatal Testing for Tay-Sachs Disease in the Light of Jewish Views Regarding Abortion." *Issues Health Care Women* 4, no. 1 (1983): 41–56.

Bleich, J. David. "Tay Sachs Disease" and "Tay Sachs Re-examined," in *Contemporary Halakhic Problems 1*. New York: Ktav, 1977, 109–115.

———. "Abortion in Halakhic Literature," in *Contemporary Halakhic Problems 1*, 325–71.

———. "Fetal Tissue Research: Jewish Tradition and Public Policy," in *Contemporary Halakhic Problems 4*. New York: Ktav, 1995, 171–202.

———. "In Vitro Fertilization, Maternal Identity and Conversion," in *Contemporary Halakhic Problems 4*, 237–272.

———. "Genetic Screening." *Tradition* 34, no. 1 (Spring 2000).

———. "Cloning: Homologous Reproduction and Jewish Law." *Tradition* 32, no. 3 (1998).

Breitowitz, Y. "What's So Bad about Human Cloning?" *Kennedy Institute of Ethics Journal* 12, no.4 (December 2002): 325–341.

———. "Halakhic Alternatives in IVF Pregnancies: A Survey." *Jewish Law Annual* 14: 29–119.

Brown, J. "Prenatal Screening in Jewish Law." *Journal of Medical Ethics* 16, no. 2 (June 1990): 75–80.

Broyd, M.J. "The Establishment of Maternity and Paternity in Jewish and American Law." *National Jewish Law Review* 3 (1988): 117–58.

Broyde, Michael J. "Cloning People and Jewish Law: A Preliminary Analysis." *Journal of Halacha and Contemporary Society* XXXIV, no. 27.

Feldman, D.M. *Marital Relations, Birth Control, and Abortion in Jewish Law.* New York: Schocken Books, 1968.

Grazi, R.V. and J.B. Wolowelsky. "Preimplantation Sex Selection and Genetic Screening in Contemporary Jewish Law and Ethics." *Journal of Assisted Reproduction and Genetics* 9, no. 4 (August 1992): 318–322.

———. "Homologous Artificial Insemination (AIH) and Gamete Intrafallopian Transfer (GIFT) in Roman Catholicism and Halakhic Judaism." *International Journal of Fertility* 38, no. 2 (March–April 1993): 75–78.

———. "Multifetal Pregnancy Reduction and Disposal of Untransplanted Embryos in Contemporary Jewish Law and Ethics." *American Journal of Obstetrics and Gynecology* 165 1991, 5 Pt 1 (November 1991): 1268–1271.

Green, R.M. "Genetic Medicine in the Perspective of Orthodox Halakhah." *Judaism* 34, no. 3 (Summer 1985): 263–277.

Halperin, M. "Human Genome Mapping: A Jewish Perspective." *Assia–Jewish Medical Ethics* 3, no. 2 (September 1998): 30–33.

Merz, B. "Matchmaking Scheme Solves Tay-Sachs Problem." *Journal of the American Medical Association* 258, no. 19 (November 20, 1987): 2636, 2639.

Mosenkis, Ari. "Genetic Screening for Breast Cancer Susceptibility: A Torah Perspective." *Journal of Halacha and Contemporary Society* XXXIV, no. 5.

Rappaport, S.A. "Genetic Engineering: Technology, Creation, and Interference." *Assia–Jewish Medical Ethics* 3, no. 1 (January 1997): 3–4.

Reichman, E. "The Halakhic Chapter of Ovarian Transplantation." *Tradition* 31, no. 3 (1998): 31–70.

Rosner, F. "The Imperative to Heal in Traditional Judaism." *Mount Sinai Journal of Medicine* 64, no. 6 (November 1997): 413–416.

———. "Recombinant DNA, Cloning, Genetic Engineering, and Judaism." *New York State Journal of Medicine* 79, no. 9 (August 1979): 1439–44.

———. "Medical Genetics—The Jewish View." *New York State Journal of Medicine* 82, no. 9 (August 1982): 1367–1372.

———. "Judaism, Genetic Screening and Genetic Therapy." *Mount Sinai Journal of Medicine* 65, nos. 5–6 (October–November 1998): 406–413.

Rosner F. and E. Reichman. "Embryonic Stem Cell Research in Jewish Law." *Journal of Halacha and Contemporary Society* 43 (Spring 2002): 49–68.

Steinberg, A. "Bioethics: Secular Philosophy, Jewish Law and Modern Medicine." *Israel Journal of Medical Sciences* 25 (1989): 404–409.

Steiner-Grossman, P. and K.L. David. "Involvement of Rabbis in Counseling and Referral for Genetic Conditions: Results of a Survey." *American Journal of Human Genetics* 53, no. 6 (December 1993): 1359–1365.

Tendler, M.D. "Alzheimer Dementia: The Judaeo-Biblical Perspective on Patient Care and Genetic Predestination or Neurocalvinism." *Alzheimer Disease and Associated Disorders* 12, Supplement no. 3 (1998): S21–S23.

Zakut, H. et al. "Chorionic Villi Sampling for Early Prenatal Diagnosis: An Option for the Jewish Orthodox Community." *Clinical Genetics* 35, no. 3 (March 1989): 174–180.

9

Islamic Perspectives on Biotechnology

Bushra Mirza

Biotechnology is unquestionably one of today's most vital and profoundly far-reaching fields. It has nearly endless applications and the potential to affect every aspect of our lives in positive ways. Biotechnological research is moving at a pace that knows no parallels in the history of biology. Yet, for most of us, modern biotechnology such as genetic engineering is new, formidably complex, and unfamiliar. This may explain somewhat why some of us find it worrisome and view it as dangerous, unethical and hence unacceptable. But one can also counter that it would be unethical not to use this powerful modern technology that bears the promise of alleviating all kinds of human suffering caused by hunger and numerous diseases. Therefore, research must proceed cautiously, potential risks need to be carefully assessed, and ethical debate must be encouraged. This ethical discussion and dialogue are all the more necessary for two reasons: The ultimate challenge in any religion or spiritual group is to be able to apply its teachings to the world in which its believers live. And issues related to biotechnology are undeniably relevant to peoples of all faiths and religions.

ISLAMIC LAWS AND GENERAL PRINCIPLES

Islam is the world's second largest religion. A conservative estimate would indicate that one-sixth to one-fifth of the world's population is Muslim. The term "Islam" means "peace and submission/surrender," and the religion of Islam stands for "a commitment to surrender one's will to the Will of God (Allah) and thus to be at peace with the Creator and with all that has been created by Him." Hence, within the Muslim world all ethical issues are examined in the light of Islamic law, known as Shariah. And the primary source of Shariah is found in the Holy Quran, which is a holy book revealed by God to the holy prophet Muhammad. The Quran gives a complete code of life, a code that incorporates economic, social, legal, and ethical principles. This code of life is further expressed in the sunnah and hadith, the traditions and sayings of the prophet Muhammad. Hadith and sunnah actually represent what prophet Muhammad said or did during his daily life that helps in explaining practical aspects of the Holy Quran. This in turn is followed by the opinions of Islamic scholars. These scholars provide their opinions with the use of analogies, that is, they rule on events that are not mentioned by the Quran and sunnah by comparing these events to similar incidents that have already been ruled upon. In this way, Islamic scholars and institutes of Islamic learning take a well-defined specific issue, examine it in detail, and decide certain questions by issuing a religious decree known as a *fatwa*.

It is very important to emphasize that in Islam there is no central institution resembling the pope or the Vatican. Therefore, juridical-ethical opinions in matters of Islamic law, or Shariah, sometimes tend to evidence a plurality of opinions, each of these based upon independent research and interpretations by legal scholars in the community.

All of this means that, when it comes to new issues like biotechnology, Muslim scholars discuss and debate specific questions on the basis of fundamental principles and guidelines. Here are some guiding principles relevant to biotechnology:

- Necessities overrule prohibition.
- Choice of the lesser of the two evils if both cannot be avoided.
- Acceptability of a deed depends upon the intention behind it.
- All things are lawful unless specifically prohibited. Similarly all things are juridically clean except those specified not to be.

Necessities overrule prohibition. To understand this rule let us take the example of food. Muslims use two key, critical terms to describe food: *halal* and *haram*. *Halal* means "permitted" or "lawful." There are no restrictions regarding consuming or using *halal* food. *Haram* means "forbidden" or "unlawful." There are prohibitions as to the consumption and use of *haram* food. Other terms used by Muslims are *makrooh* and *mashbooh*. *Makrooh* means "religiously discouraged" or disliked. *Mashbooh* means "suspected."

Eating pig is forbidden, haram. However, treating diabetic patients with insulin that is obtained from a pig source is permissible because of necessity, given that no alternative is available. Similarly alcohol is also forbidden, or haram. But until alcohol-free medicines can be prepared, there is no prohibition against using medicines containing alcohol.

Choice of the lesser of the two evils if both cannot be avoided. Violating the human body, whether living or dead, also violates the teachings of Islam. Therefore, one might claim that organ transplantation is not permissible because that requires incising the body of a living donor or of a cadaver and obtaining the organ to be donated. Nevertheless, the saving of life is a necessity that carries more weight than preserving the integrity of the body of donor or cadaver. Hence, organ donation and transplantation are sanctioned since the injury to the body of the donor or cadaver is a lesser evil compared to letting a patient die.

Acceptability of a deed depends upon the intention behind it. Islamic teachings place a great deal of emphasis upon *neeyat*, meaning "intention." Muslims believe that any reward for deeds rests upon the intentions behind the deed. Therefore, whether an action is sometimes permissible or not depends primarily upon the intention behind the deed. For example, according to Islamic teachings, euthanasia is absolutely unacceptable and unlawful. A patient should receive whatever medical care is possible, and this includes material and emotional support from family and friends. Human life per se is to be valued unconditionally, irrespective of circumstances. The idea of a "life not worth living" does not exist in Islam. This means that the taking of life *in order to* end life is not acceptable. Yet suppose that a person dies because he or she is given a lethal dose of a painkiller, and this dose is administered *in order to* alleviate the person's suffering, not to take that person's life. In this case, Islamic teachings would view this as acceptable.

All things are lawful unless specifically prohibited. Similarly all things are juridically clean except those specified not to be. Islamic teachings clearly designate foods or actions that are forbidden. However, we need to be very careful in how we interpret and understand this. The prohibition of a food or drink does not necessarily mean that it is juridically unclean. For instance, even though drinking and handling alcohol is forbidden because it is an intoxicant, there is no problem with using perfumes or cologne in which alcohol is used as a solvent for fragrance and aroma, or in using creams that also contain alcohol.

Now that we've looked at some of the basic rules and principles of Islamic laws, let us proceed further and discuss issues in specific areas in biotechnology: genetically modified food, assisted reproductive technology, prenatal testing, germ cell research, gene therapy, and cloning.

GENETICALLY MODIFIED FOOD

Biotechnology has abundant applications for the food industry. In general, outside of a few reservations about products that need to be reviewed on a case-by-case basis, Muslims accept biotechnology ingredients and enzyme cultures. However, the Muslim community is also aware that it is sometimes rather difficult to make halal determinations (that is, to claim that food is permissible) of food ingredients, food products, and modified species of animals and plants produced through biotechnology. It is thus important that information should be made available to Muslims concerning the theories, concepts, and practices in food biotechnology and genetic engineering so that they can more properly evaluate their halal nature (permissibility) or otherwise.

Muslims were initially concerned that genetically modified (GM) food might contain genes from animals, whereas Muslims have no problem in using GM food crops that have so far been approved and developed with genes from plants. The U.S. Islamic Jurisprudence Council (IJC) has declared all current GM foods on the market to be halal and therefore fit for consumption by Muslims. It is important to mention here that according to Islamic law, transformation, that is, the process by which an object changes into another, totally different object with different properties, can also turn what is deemed to be unclean into a clean object. Therefore a substance may be extracted from a prohibited source. Yet by going through this transformation, via chemical or other changes that modify its nature, the end product is considered permissible by the Shariah, or Islamic law. For example, soap that is produced by treating and transforming pig fat or fat obtained from a dead animal (by themselves unclean and prohibited sources) becomes a clean compound through the process of transformation. Therefore it is permitted to use this soap. Similarly, Muslims have no objection to the use of bioengineered chymosin (rennin) in the production of cheese. Furthermore, Muslim leaders have ruled that simple gene additions that lead to one or a few new components in a species are acceptable. However, they have not yet resolved whether or not a gene that is derived from swine (a prohibited animal) constitutes an exception to this acceptance. Neither has it been determined about the status of more significant changes in the genetic makeup of species.

ASSISTED REPRODUCTIVE TECHNOLOGY

Assisted reproductive technologies (ARTs) make up one of the fastest growing areas in medicine. While this technology has certainly improved the chances for an infertile couple to conceive a child, it has evoked strong opposition from religious groups as well as human rights activists who object on legal and moral grounds. Islam is not opposed to treating infertility, and infertile couples seeking treatment for their infertility are not viewed as violating Islamic law. In fact, seeking treatment for infertility is encouraged since it aims for the good of procreation. However, though the aim is good, the means to achieve it that involves ARTs elicits debate.

Note the steps involved in a cycle of treatment such as in vitro fertilization (IVF). First, there is drug stimulation of the ovaries in order to produce multiple follicles. This is done in order to collect oocytes. The oocytes are then inseminated with sperm and left to fertilize. The embryos are then cultured for a period of two to three days, after which they are selected based on morphology and cell counts before being transferred back to the mother's uterus. Where infertility is due to the husband's condition of being either azoospermic (no sperm) or some other male factor, an alternative would be to inseminate the oocytes with the sperm from a donor. If infertility is due to the woman, embryos are implanted into the womb of a surrogate mother.

From an Islamic perspective, IVF is permissible only under certain conditions. The woman's oocyte may only be fertilized with her husband's sperm and only if they are still married. There must not be a third party such as an anonymous donor or a surrogate mother. Islamic teachings absolutely prohibit insemination of the woman's egg with the sperm of a man who is not her husband. Islamic scholars have addressed this issue of using sperm donors and have declared it comparable to adultery. They therefore regard it as a grievous crime and a great sin. Similarly surrogacy is entirely illegitimate. As the Holy Quran states, "None can be their mother except those who gave birth." And surrogacy creates confusion about who is the mother of the child as well as involves a woman who is not married to the father of the child. However, since polygamy is permissible, some jurists do permit in vitro fertilization between the husband's sperm and an egg from a legally married wife that is then implanted into the legally married second wife. Although this involves taking eggs from one woman and implanting it in the uterus of the other woman, both of these women are still within the same marriage contract with the man whose sperm is used for fertilization.

GENETIC SCREENING AND SELECTION OF THE EMBRYO

Genetic screening enables us to identify carriers of genetic disorders. It can be used to detect genetic disorders of the fetuses, through prenatal testing, or of the adults. The information obtained through this procedure can be used for genetic counseling or to protect the lives of the mother and the fetus. Genetic counselors have occasionally recommended the abortion of severely afflicted fetuses. In a more advanced technique called preimplantation genetic diagnosis (PGD), genetic screening of an embryo is done during assisted reproduction before its implantation into the uterus of a woman. PGD involves sampling one of the blastomeres that is present in the cleaving two- to three-day-old embryo and screening its genetic structure. The purpose is to screen for genetic defects that may be present in the embryo. Any embryo that does not carry genes for genetic diseases is then implanted into the uterus.

From an Islamic point of view, the genetic screening of populations at risk in order to detect genetic abnormalities and for the purposes of genetic counseling and curing birth defects is acceptable. Similarly, prenatal diagnosis in order to protect the mother's life or health is also permissible. Islamic teachings do not permit the selection of an embryo on the basis of genetic structure through PGD. Furthermore, Islamic teachings forbid aborting a fetus for reasons of genetic abnormality. Muslims must accept all the decisions and gifts of God (Allah), and they must not determine who deserves to live for however long based upon their own desires. Spiritual maturity in Islam requires prohibiting any claims of superiority of one person over the other. All persons are equal. In the Holy Quran, the only valid claim to nobility rests upon being God-fearing. Therefore, deliberately selecting an embryo interferes with the work of the Creator. Selecting for gender is also strictly prohibited. As the Holy Quran states, "He [Allah] creates what He wills, He bestows females upon whom He wills, and He bestows males upon whom He wills. For He is all knowledgeable and all powerful."

FATE OF THE SPARE EMBRYOS

A procedure like IVF often results in the availability of numerous spare oocytes and embryos that are not transferred to the uterus of the mother. From the Islamic perspective, the fate of these remaining embryos is a debatable issue. Patients sometimes consent to discard the spare embryos or to freeze them (cryopreservation). Freezing embryos has it advantages. For one, if a woman decides to undergo IVF again, she does not have to repeat the drug stimulation cycle and can thus avoid having to experience any side effects from the stimulant drugs. Cryopreservation techniques can store embryos up to a few years, so that if the mother later decides to have a child, these embryos can be thawed and implanted in her uterus. Muslim scholars consider this process legitimate so long as the woman from whom the oocytes were obtained is still within the marriage contract with the husband whose sperm was used to fertilize her own oocytes.

Spare embryos can also be donated to a childless couple. Islamic teachings consider this illegitimate because it involves a third party to which the husband was not legally married and thus violates Islamic laws. Surrogate motherhood also produces its fair share of social and psychological complications.

Spare embryos can be donated for research purposes. Embryo research can further improve our knowledge about assisted reproduction, the diagnosis and prevention of genetic disease, and the development of better therapies for birth defects. In the light of these benefits, Muslim scholars have differing opinions on this issue. Some Muslim scholars believe that since a human being is God's creation, once that human being comes to life, it should not be destroyed or manipulated. Other scholars allow embryo research for therapeutic purposes as long as there is the prior consent of the couple undergoing infertility treatment. At the same time, these scholars add that the embryos that have been researched upon must not be transferred back to the uterus of the mother or to another woman.

This controversy stems from the following question: If an embryo formed after artificial fertilization is not yet in the womb of its mother within a few days, should it be considered a human being, with all the rights of a human being? According to the Shariah, we should make a distinction between actual life and potential life. Indeed an embryo is valuable. It has the potential to grow into a human being, but it is not yet a human being. Destroying such an embryo is not called abortion. In the Quran as well as in hadith, different stages of fetal development have been described, and the stage for "ensoulment" is indicated at 120 days. Muslim jurists have made a clear distinction between the early stages of pregnancy (the first forty days) and its later stages. For instance, it is mentioned that if someone attacks a pregnant woman and aborts her baby in the early stages of her pregnancy, that person's punishment will be less than that of a person who commits this act during full pregnancy. And if he kills that child after its birth then he is liable to be punished for homicide. So there is nothing wrong in doing research especially if this research has a potential to cure diseases.

However, it is important to keep in mind that Muslims have strong reservations against the misuse of embryos. Embryo research surely has the potential for misuse. For instance, with respect to donors of cells, we should anticipate various possible misuses and establish safeguards against them. Possible misuses might include physicians requiring infertility patients to go through extra cycles of ovulation purely for the purpose of obtaining more embryos. Or they may themselves pay women to produce embryos. Or they may cultivate embryos without the consent of the donor. And in setting up safeguards, authorities should also clarify the difference between, on the one hand, the use of spare embryos from in vitro fertilization procedures that would be destroyed regardless and, on the other hand, the deliberate production of embryos for stem cell research.

Each year thousands of embryos are wasted in fertility clinics worldwide. Such embryos should not be wasted; they should be used for research. If embryonic stem cell research has potential to relieve human disease and suffering, then not only is it permitted but it is *obligatory* to pursue this research. However, this research must only involve those embryos that were created for the purpose of in vitro fertilization and that would otherwise be destroyed.

GENE THERAPY

A most noteworthy outcome of the Human Genome Project lies in its application to gene therapy. Gene therapy essentially means replacing defective genes with healthy genes. This can be done with somatic cells or germ cells. Somatic gene therapy simply relieves the patient from disease symptoms while germ line gene therapy more radically modifies the genetic makeup of subsequent generations. According to a Muslim perspective, human gene therapy should be limited to therapeutic indications. Muslim scholars encourage somatic cell gene therapy since it leads to the alleviation of human suffering. Yet Muslim scholars also forbid germ line gene therapy because of its consequences such as the attempt to enhance a particular characteristic like intelligence or beauty. Genetic engineering aiming for such enhancement is absolutely prohibited. It deliberately interferes with God's creation and design, and this, in turn, may lead to an imbalance within the entire universe.

HUMAN CLONING

Cloning entails making identical genetic copies of an individual outside of the normal process of sexual reproduction. Cloning techniques have already been used for quite some time with plants in order to reproduce plants with desired characteristics. Various animal species have also been reproduced via cloning techniques. However, human cloning remains highly controversial. Muslims are especially concerned about how human cloning would impact upon the fundamental relationship between man and woman. They are also worried that human cloning would adversely affect certain life-giving aspects of spousal relations such as parental love and concern for their offspring. According to Islamic teachings, interpersonal relationships are central to human life. The prophet Muhammad is reported to have said that nine-tenths of religion pertains to human-to-human relationships whereas only one-tenth deals with God-to-human. The Holy Quran acknowledges the sex pairing between man and woman as a universal law; furthermore, reproduction is only allowed within the marital relationship.

The general opinion in the Muslim world today is that cloning research that deals with laboratory animals is encouraged since such research would add to our understanding of biological processes and would contribute to human well-being. After sufficient animal studies and research to order to ensure their safety, cloning techniques may then be applied in humans in order to alleviate human sufferings and bring about effective treatment of diseases. However, under no circumstances whatsoever should cloning be applied in order to produce a genetically identical human individual.

CONCLUSION

To summarize, Islamic teachings hold that the applications of biotechnology are not only permissible but also obligatory if they result in either alleviating human suffering or in saving human life. According to the Holy Quran, "whoever saves a life, it would be as if he saved the life of all the people." In this light, any opposition to biotechnology does not at all mean that Islam is opposed to technological progress. Rather, Muslim scholars seek to examine and understand all aspects of particular biotechnological applications to ensure that these applications are consistent with the basic teachings and fundamental principles of Islam. It goes without saying that any serious discussion of the moral justifiability of biotechnology must consider all kinds of potential risks such as political abuse, commercial exploitation, and adverse effects upon interpersonal relationships. Moreover, Islam is a flexible religion and acknowledges the need to accommodate its teachings to life's realities and necessities and to human well-being. However, in attempting to achieve these goals we should not contradict the teachings of the Holy Quran and the guidelines of Islamic laws.

NOTES

The author is thankful to Dr. Anwar Nasim, Science Advisor, COMSTECH, Islamabad, for his critical reading of the manuscript and helpful suggestions.

BIBLIOGRAPHY

Ahmad, N.H. "Assisted Reproduction—Islamic Views on the Science of Procreation." *Eubios Journal of Asian and International Bioethics* 13 (2003): 59–61.
Al-Qarawadi, Y. *The Lawful and the Prohibited in Islam, Al-Halal wal Haram fil Islam*. Kuala Lumpur, Malaysia: Islamic Book Trust, 1995.
Chaudry, M.M. and J.M. Regenstein. "Implications of Biotechnology and Genetic Engineering for Kosher and Halal Foods." *Trends in Food Science & Technology* 5 (1994): 165–168.
Eskandarani, H. *Assisted Reproductive Technology: State of the ART*. Publications of the Islamic Educational Scientific and Cultural Organisation (ISESCO). Saudi Arabia: Alwafa Printing Press, 1996.
Nasim A. "Ethical Issues: An Islamic Perspective." Workshop on Asian Genomics: Cultural Values and Bioethical Practices. Leiden, the Netherlands, 28–29 March 2002.
Nasim A. "Ethics of Stem Cell Research: An Islamic Perspective." Third International Conference of Bioethics: Ethical, Legal and Social Issues in Human Pluri-potent Stem Cell Experimentation. Graduate Institute of Philosophy, National Central University, Chungli, Taiwan, 2002.

Nasim, A. "Ethical Issues of the Human Genome Project: An Islamic Perspective." *Bioethics in Asia: Proceedings of the UNESCO Asian Bioethics Conference, 3–8 November 1997.* Kobe and Fukui, Japan, 1997, 209–214.

Recommendations of the Eighth Fiqh-Medical Seminar. Al-Marzuq Centre for Islamic Medical Science, Kuwait, 22–24 May 1995, www.islamset.com/bioethics/8thfiqh.html.

Riaz, M.N. "Halal Food—An Insight into a Growing Food Industry Segment," www.icbcs.org/halal.htm.

Sachedina, A. "Testimony of Abdulaziz Sachedina, University of Virginia," in *Ethical Issues in Human Stem Cell Research* 3. Rockville, Maryland, 2000, G1–6.

Sachedina, A. "Islamic Perspectives on Cloning," www.nooralnissa.hpg.ig.com.br/islamicperspectivesoncloning.htm.

Serour, G.I. "Reproduction Choice: A Muslim Perspective." In *The Future of Human Reproduction: Ethics, Choice and Regulation*, edited by J. Harris and S. Holm. Oxford: Clarendon Press, 1998.

Serour,G.I. *Ethical Implications of Human Embryo Research.* Publication of the Islamic Education, Scientific and Cultural Organiztion—ISESCO, 2000.

Shinwari, Z.K. and A. Nasim. "Bioethics and Pakistan: An Islamic Perspective." International Conference on Dialogue and Promotion of Bioethics in Asia. Singapore, 9–10 March 2003.

Siddiqi, M. "An Islamic Perspective on Stem Cell Research." 2002, www.islamicity.com.

10

Agricultural Biotechnology in African Countries

Martin O. Makinde

Biotechnology's promise to revolutionize the world in both industrial and agricultural practices has become a reality, and Africa must not again be left behind. In brief, biotechnology is the manipulation and use of biological organisms to make products that benefit human beings. Actually, biotechnology is as old as humanity, with numerous ancient records of *traditional* biotechnologies such as making bread, beer, wine, cheese, and yogurt. Later *conventional* biotechnologies include improved food production through genetic modification (GM) in the form of selection, breeding, and mutation through radiation or chemical methods. *Modern* biotechnology techniques such as DNA forensics, tissue culture, molecular breeding, genomics (study of an organism's genes), proteomics (study of proteins), and metabolomics (study of metabolic pathways) are now in use and lead to new products and discoveries. There is no controversy over traditional and conventional, first and second generation, biotechnologies. And while much public attention and controversy in modern, third-generation biotechnology focuses more on gene transfer and cloning, not all gene transfer is controversial. GM food-processing enzymes, nutrition additives, medicines, and even industrial enzymes—used daily by most of us—evoke little debate. Furthermore, not all GM plants are controversial. What poses the most heated dispute are the GM food groups.

THE LEADING ROLE OF AGRICULTURE

Africa has 13 percent of the world's population. Its steadily high growth rate continues to place an enormous burden on its economic growth by increasing food insecurity and reducing environmental sustainability. According to the 1997–1999 Food and Agriculture Organization of the United Nations (FAO) reports, about 200 million people, 28 percent of Africa's population, are chronically hungry compared to 173 million ten years earlier. Of these, some forty million children are severely underweight. At the end of the 1990s, thirty African countries had more than 20 percent of their populations undernourished, and eighteen of these countries recorded more than 35 percent of their populations as being chronically hungry. In 2001, about twenty-eight million people in Africa were facing food emergencies due to droughts, floods, and civil wars. Of these, some twenty-five million people needed emergency food aid and agricultural assistance. More than fifty million Africans, mostly children, suffer from vitamin A deficiency while 65 percent of African women of childbearing age are anemic. In 2000, the World Food Programme spent an estimated $18.7 billion on food aid.[1]

Hunger is a critical issue particularly in sub-Saharan Africa. The major causes of food insecurity in sub-Saharan African that called for exceptional food emergencies included:

- civil strife—Democratic Republic of Congo, Sierra Leone, Sudan, Somalia, Liberia, and parts of Uganda
- population displacement—Angola, Guinea-Bissau, and Sierra Leone
- below-normal production—Rwanda and Burundi
- economic sanctions—Burundi
- refugee issues—DR Congo
- large numbers of vulnerable people—Ethiopia and Rwanda
- localized weather adversities—Ethiopia and Zambia
- impact of past civil strife—Liberia and Mozambique
- shortage of farm inputs—Liberia
- localized deficits—Mauritania and Sudan
- poor harvests—Somalia and Zambia
- postharvest losses—Zambia[2]

We can also add to this all sorts of environmental constraints in the different developing countries. We can identify four broad agro-ecological zones: humid and peri-humid lowlands; hill and mountain areas; irrigated and naturally flooded areas; and drylands and areas of uncertain rainfall. Within each of these agro-ecological zones, there is a wide range of farming systems as well as a blend of traditional and modern production systems.

It is estimated that the world food production is adequate to feed all its inhabitants. Nevertheless, a 1995–1997 study disclosed that in developing countries there were roughly 790 million undernourished people, that is, people whose food intake was insufficient to meet basic energy requirements on a continuing basis.[3] Hunger and poverty are thus endemic in

Africa, and both are influenced and determined by many different demographic, environmental, economic, social, and political factors.

Increasing food demands clearly requires more effective food production. Yet Africa faces formidable obstacles. What are the greatest constraints to agricultural production in Africa? They include drought, insufficient resources to purchase external outputs, floods, lack of diversified food products, pests, and diseases. Other factors include: widespread conflicts; deterioration of trade terms for non-oil primary commodities; Africa's vulnerability to trade declines due to its failure to diversify into more dynamic product lines; faulty economic policies and poor governance that create a hostile environment for investments and growth and result in a dramatic decline in both the amount and efficiency of public and private investments; lack of investments in people and in the basic infrastructure; and the AIDS pandemic.

Consider farmers who, in most rural areas of the developing countries of Africa, are subsistence farmers. On these farms, production depends essentially upon family units that offer relatively little for sale and obtain little, if any, goods or services from outside sources. Such family units may then have to depend on financial support from outside nonfarming relations. These farmers grow plants and keep livestock in order to satisfy the needs of their families rather than to meet the demands of a market. In fact, in many parts of Africa, those animals that are sold happen to be the oldest and weakest, not younger and more vigorous since the security value of the latter is greater.[4]

One solution to poor nutrition requires expanding food production within Africa itself. Now it is well-known that most of the world's food consumption takes place in the countries in which the food is produced. This means that in order for Africa to meet 85 percent of its food self-sufficiency by 2015, it needs an increased output of 118 million tons of its projected need of 139 million tons of cereals.[5] Until this is achieved so that the incidence of hunger is reduced along with the costs of importing food supplies (food insecurity is the highest in sub-Saharan Africa), there is little prospect of attaining the high rates of economic growth to which the New Partnership for Africa's Development (NEPAD) aspires. In any case, agriculture remains crucial for economic growth in most African countries. Agriculture is the engine for overall economic growth in Africa for these reasons:

- it provides 60 percent of all employment
- it constitutes the backbone of most African economies
- it is the largest contributor to GDP
- it is the biggest source of foreign exchange
- it accounts for about 40 percent of the continent's hard currency earnings
- it is the main generator of savings and tax revenue
- it is the dominant provider of industrial raw material

THE GREEN REVOLUTION AND ITS IMPACT ON AFRICA

The Green Revolution started in Mexico in the 1940s with the use of genetic breeding techniques aimed at improving the yields of basic food crops such as maize and wheat. In 1948, Mexican farmers planted 1,400 tons of improved maize seed and did not need to import maize. Improved breeding programs spread to South America, India, and Africa resulting in similar yield increases with other crops.[6] However, such success required that crops be supplemented with fertilizers and optimal supplies of water. For most African farmers, both of these were and are still in short supply. The Green Revolution, therefore, was least successful in sub-Saharan Africa.[7] Following this, scientists and farmers realized that they needed to forge partnerships in order to develop alternatives to inorganic fertilizers and pesticides, to improve soil and water management, and to enhance earning opportunities for the poor, especially women. It was further suggested that "genetic engineering has a special value for agricultural production in developing countries. It has the potential for creating new plant varieties—not only to deliver higher yields but also to contain the internal solutions to biotic and abiotic challenges, reducing the need for chemical inputs, such as fungicides and pesticides, and increasing tolerance to drought, salinity, chemical toxicity and other adverse conditions. Genetic engineering can achieve not only increasing productivity but also higher levels of stability and sustainability."[8]

A COMPELLING CASE FOR GM CROPS IN AFRICA

The biotechnology industry is a multibillion dollar business with ventures in human health care, industrial processing, environmental bioremediation, food, and agriculture. These technologies reflect market realities and are used primarily to provide products for the developed countries. Yet they are appropriate for food production and agriculture in developing countries as well. Agricultural biotechnology is indeed one of the most powerful tools for agricultural development for numerous reasons. It is a vital sustainable development tool that can address the food security issues in Africa. It has the far-reaching potential to:

- enhance food security and hence reduce poverty and hunger through increased agricultural productivity and increased yield per unit area from drought-tolerant and disease-resistant transgenic crops
- process products and add value to domestic products
- improve nutrition through biofortification of crops using biotechnology methods
- reduce environmental pollution through decreasing the use of pesticides

- improve health delivery systems such as vaccine production through molecular-marker-assisted-breeding and DNA fingerprinting
- bring about cost-effective labor through the use of herbicide-resistant crops and less reliance on spraying and weeding cycles
- make a positive economic impact due to reduced cost of inputs
- address Africa's specific agricultural needs and environmental conditions
- increase stability of crop production following the development of drought-resistant, virus-resistant, and insect-resistant seed

Along these lines, recent studies have shown that for African countries, a 10 percent increase in yield led to a 1.6 percent gain, resulting in a 9 percent decrease in the percentage of those living on less than $1 per day.[9] Biotechnology thus offers hope that Africa can effectively address its critical production and nutrition constraints. Biotechnologies can help develop crops that are resistant to nonbiotic stresses (acid, alkaline soils, salinity tolerance, and pests). Biotechnology can aid in the production of disease-free planting material (e.g., banana, plantain, cassava). It can reduce the need for fertilizers and pesticides. It can improve livestock productivity through disease resistant stocks and cheaper vaccines. And biotechnology can improve the nutritional quality of food crops.

Up to now, biotechnology research has been concentrated in the developed countries with fewer than 20 percent of the trials in the developing countries. Land that has been cultivated with genetically modified crops has grown from two million hectares in 1996 to more than fifty-eight million in 2002.[10] Among the developing countries, Argentina, China, Mexico, and South Africa have already begun significant commercial planting of GM crops. And Egypt, Kenya, and Zimbabwe have also initiated commercial applications of biotechnology.

Biosafety Status

There are certainly biosafety concerns, and the most prominent is the fear that living modified organisms may have adverse effects upon the conservation and sustainable use of biodiversity, especially in instances involving transboundary (cross-border) movement. Concerns are related to unproven socioeconomic and ecological benefits and risks that these technologies may bring. As a result, governments worldwide signed the biosafety protocol in January 2000 to ensure safer transfer, handling, and use of living modified organisms resulting from modern biotechnology. To date, eleven African countries have ratified the Cartegena Protocol on Biosafety: Botswana, Djibouti, Kenya, Lesotho, Liberia, Malawi, Mali, Mauritius, Mozambique, Tunisia, and Uganda. South Africa, Zimbabwe, and Malawi have established GM legislation and functioning frameworks. Meanwhile, those with draft legislation and interim frameworks are Egypt, Kenya, and Uganda. And those with draft legislation and frameworks yet to be revised are Cameroon, Côte d'Ivore, Mauritius, Namibia, and Zambia. Furthermore, forty-three African countries have United Nations Environment Program-Global Environment Facility (UNEP-GEF) biosafety development processes.

Despite biosafety concerns, there is an increasing awareness of biotechnology's positive potential, and other countries such as Nigeria, Côte d'Ivoire, and Malawi have expressed strong interests in biotechnology. There are already several success stories whereby biotechnology has increased crops' resistance to biotic and nonbiotic stresses, reduced the cost of pest controls, and created new employment opportunities. For instance, there is the wide adoption of disease-free banana plantlets in Kenya, the use of pest-resistant cotton varieties in South Africa, and the use of new vaccines in Kenya and Zimbabwe. Most African countries are involved primarily in second-generation plant tissue cultures involving crops like cassava, yam, pineapple, cocoa, bananas, and cowpeas. And South Africa is the only African country to commercialize the following GM crops: cotton, yellow maize, soybeans, and white maize.

African Crops and Diseases

The following crops are specific to Africa: cassava, bananas, sweet potatoes, millet, sorghum, and maize. Cassava is a staple food for many in the continent and provides a lot of calories. The tuber can be boiled, fried, powdered, or fermented. While plant breeders have succeeded in increasing the maturity, size, and number of edible tubers, they are often infected by cassava mosaic virus. Scientists have produced virus-resistant varieties, yet these have not been commercialized.[11] Another problem with this tuber is its high levels of cyanide and lethal effects if not properly processed. This can also be genetically modified.

Bananas are the world's fourth most important crop, providing more than one-quarter of all food calories. There are efforts to introduce genes that encode resistance to the most serious fungal diseases.

Sweet potatoes in Africa are mainly grown by resource-poor women. Sweet potatoes are important for food security as they yield more food energy and micronutrients per unit area than any other crop. Scientists in South Africa, Kenya, and Uganda have succeeded in improving the protein content of sweet potatoes by a factor of four, from the traditional 3 percent to 12 percent with a potential significant effect on Africans.[12] However, viruses and weevils cause significant yield losses. Now there is a joint project between Monsanto, the International Service for the Acquisition of Agri-biotech Applications (ISAAA), the Kenya Agricultural Research Institute (KARI), and the

International Potato Centre to address these problems using genetic engineering. This is a royalty-free agreement between Monsanto and KARI and other interested African countries. It is expected that African farmers will receive transgenic virus-resistant sweet potatoes and weevil-resistant varieties by 2004.[13]

Maize streak virus is an endemic virus that causes havoc in maize crops in Africa. Scientists in South Africa and the United Kingdom are working on alternative ways of developing resistance to this devastating disease that affects this crop that many Africans love to eat three times a day.[14]

How Can Africa Realize These Biotechnology Needs?

In order to realize the benefits of biotechnology, Africa needs to develop mechanisms that promote technology development, transfer, and diffusion. Africa also needs to create research centers of excellence in African staple crops. Experts must make a concerted effort to develop effective science and technology institutions as well as research activities capable of developing and adapting to world-class technologies. What do such research activities involve? To begin with, these activities include capacity building in human resources and in the infrastructure. This capacity building in both human and material resources is critical to the sound application and management of biotechnology and can be addressed by encouraging collaboration among the various African stakeholders so as to enhance biotechnology activities on the continent. And because the lack of capacity poses a key problem in Africa, governments and international organizations need to form partnerships to narrow the knowledge and technology divide between the developed and developing worlds.

Second, Africa needs to develop effective communication and information-sharing systems. Public perception remains a critical factor, especially among the literate. At present, outside forces are pressuring Africa to reject biotechnology. It is therefore vital to provide consumers, media, and decision makers with accurate and objective information on biotechnology to ensure that biotechnology is represented fairly and at all levels of African society.

Finally, Africa needs to establish appropriate biopolicy strategies through legislation and regulatory and administrative structures. African governments need to formulate relevant and appropriate biotechnology and biosafety policies that take into account the specific needs of Africa. This includes creating an environment that promotes and facilitates biotechnology research and development in agriculture, health, industry, and environmental management. Of course, one of the keys to all of this is to establish funding mechanisms that support biotechnology. In turn, biotechnology policies need to be based on clearly articulated national priorities and goals.

BIOTECHNOLOGY PATENTING AS A KEY CONSTRAINT

Intellectual property rights (IPR) is a broad term for the various rights protected by the law for the economic investment in creative effort. The principal categories of intellectual property relevant to agricultural research are patents, plant variety rights, trade secrets, and trademarks.[15] The 1993 Convention on Biological Diversity (CBD) and the Trade-Related Aspects of Intellectual Property Rights (TRIPS) agreement of 1994 compose the two main international agreements on IPR. The TRIPS agreement made it obligatory to patent the inventions of products and processes in all fields of technology for a minimum period of twenty years. However, there are some exceptions for plants and animals and the biological processes that produce them. Nevertheless all other products, such as agrochemicals, are subject, according to this agreement, to patent protection. Biotechnological innovations arising from microorganisms and microbiological processes are also eligible for patent protection under the TRIPS agreement. As a result, an increasing number of developing countries are revising or setting their own systems to protect IPR despite earlier opposition to it in the 1960s. The World Trade Organization (WTO) requires its members to introduce international standards for the protection of IPR.

One of the most controversial issues related to IPR is the practice in which companies from developed countries patent genetic material from developing countries, often without the knowledge and consent of the owners of such resources in the developing countries. Moreover, there is an apparent lack of adequate mechanisms to prevent this from happening. This gives rise to an ongoing failure to both acknowledge the contribution that farmers in developing countries have made and to share any benefits with them.[16]

Biotechnology companies have used universities to gain easier access to valuable biological material. This occurs through research collaborations between institutions in developed and developing countries, a collaboration resulting in researchers acting merely as sample collectors. Critics point out that this collecting of biological material is often the main motive behind the collaboration. Nevertheless, the concept of the IPR is justified as a reward system for innovation, capital risk-taking, and simple hard work, where the beneficiary is not only the patent holder but includes farmers, consumers, and the entire community. However, this issue of patents regarding genetic material obtained from developing countries being granted to companies in the developed countries is too important to be ignored. It is gratifying to note that two well-known seed companies operating in Africa have come up with royalty-free policies for some GM crops under field trials for African farmers that include sweet potatoes, cassava, and banana.

The International Treaty on Plant Genetic Resources for Food and Agriculture recognizes the rights of farmers to save, use, exchange, and sell farm-saved seeds and other propagating material as well as their right to share in the benefits arising from plant genetic resources. These are fundamental to the realization of farmers' rights. We of course need to recognize these rights as well as hold to a system whereby private entities are rewarded for their research and development efforts. But this does not preclude farmers from exercising their rights in relation to genetic material that they developed and nurtured in the form of farmer varieties and land races (unimproved varieties). A system of farmers' rights for plant varieties will have to resolve questions as to proof of human intervention, ownership, and the ability to describe such varieties botanically. National governments should be encouraged to support the collection, conservation, and maintenance of such biodiversity in situ and ex situ.

ETHICAL ISSUES

The most prominent ethical issue in relation to transgenic plants is the belief that GM plants are "unnatural." In this light, the public needs to be made aware that actually no plants by themselves are natural, all being the result of extensive genetic modifications through thousands of years of selection and breeding by farmers. Molecular genetics underscores this fact that plant genes have evolved from other genes. Indeed, various techniques of transferring or altering genes used in traditional agriculture—cell fusion, artificial pollination, embryo rescue, irradiation for mutagenesis, and classical breeding—are, as with gene cloning, ways of manipulating nature for the benefit of human beings.[17] What is truly of ethical importance to Africa is whether it is ethical to starve people of food, including GM food aid, based on nonscientific evidence.

Negative Perceptions Created by Active Antibiotechnology Campaigners

Within the past few years, Africa has become the locus for antibiotechnology activists. They have taken their campaign to new heights, advising African governments on issues such as food aid, access, benefit sharing of genetic resources, and the possible development of unworkable biosafety frameworks. It is thus all the more critical for African governments to be provided with the facts of agricultural biotechnology and its products through structures such as Africa Union, the New Partnership for Africa's Development (NEPAD), Forum for Agricultural Research in Africa (FARA), regional organizations like the Economic Community of West African States (ECOWAS), and the South African Development Community (SADC) and by

African scientists, farmers, consumer organizations, and nongovernmental organizations (NGOs) like AfricaBio and African Biotechnology Stakeholders Forum (ABSF).

SOCIOECONOMIC ISSUES

In most if not all African countries, agriculture accounts for a large share of total income and employment. Boosting growth in agricultural productivity will therefore have a direct effect upon income, either through increasing own-farm production or through providing greater agricultural employment and income opportunities for small holders and landless laborers. This has been well demonstrated by the small-scale *Bt* cotton farmers in South Africa.

In addition, studies have shown that every percent increase in agricultural income generates an equal percent increase in the nonfarm income. It is estimated that the rural nonfarm economy accounts on average for at least twenty percent of full-time rural employment in Africa, and nonfarm income for about 40 percent of rural incomes. Furthermore, the impact of increased agricultural productivity is not restricted to the rural sector. Consumers, especially poor urban dwellers who spend a large proportion of their income on food, benefit directly from cheaper food prices. This has a positive impact on the competitiveness and growth of other sectors of the economy.

TRADE ISSUES

Africa may be reluctant to explore the possible benefits of using GM crops because of the implications of European Union (EU) regulatory policies for GM crops. The challenges are on the GM-specific regulations, all commodity shipments that contain GM varieties (above certain thresholds) are considered GM. Therefore, non-GM products might be labeled as GM. This means that a large proportion of the exports of GM crops from the main producing countries are considered GM for regulatory purposes. As a result GM-specific regulations have created two categories per crop in the commodity trade where previously there was only one. In addition, most African countries are yet to have the biosafety policies in place. Also, countries have very different domestic regulations of GM products. U.S. regulators consider GM foods as not substantially different from the conventional variety (i.e., they are substantially equivalent). The EU adheres to the "precautionary principle." Regulators in the EU have GM and conventional varieties as different products because of the perceived risks. The commodities are overregulated and segmented, especially by Africa's main trading partners. It may be high time that Africa started looking inward and have better penetration of its huge market like China and India.

CONCLUSION

The economic benefits of GM crops are certainly tangible, and Africa's poorest farmers should have access to this technology. If the developing world's legitimate demands regarding food production are not attended to, there will be dire consequences for both developing and developed worlds.

As scientific and environmental elephants butt heads, the hopes of Africans should not be trampled upon. For the benefit of humankind, we must end the squabbling over biotechnology and allow objectivity to prevail.[18] Scientists are developing strains of rice, cassava, and other staple foods that are fortified with more nutrients. Crops are also being developed to generate their own protection against pests and diseases. Of course there may be potential risks. But the benefits and the severity of the need make it clear that the sensible approach is to minimize and manage the risks, not to abandon biotechnology.

In conclusion, African governments and scientists have endorsed this tool for agricultural production. Therefore we need to be proactive and collaborate in the following areas. First, we need to set research and development priority settings, especially for indigenous African crops and also where agriculture and health care merge. Biotechnology research should be used only to solve specific problems where it has comparative advantage. In addition to technical considerations, priority setting should take into account national development policies, private sector interests, and market possibilities. We need to work on capacity building and increased technology transfer, information sharing, networking, and business development. Truly productive research requires a critical mass of expertise and facilities. Biotechnology research entails well-equipped laboratories with proper working conditions, a constant supply of good quality water, a reliable power supply, and organized institutional support. We need to construct effective policies. Strong partnerships among African countries will be needed in order to effectively implement risk assessment, risk management, and risk communication. Along these lines, South Africa has a leading role to play as it is one of the few developing countries to successfully commercialize GM crops. This experience has benefited both small-scale and emerging farmers. South Africa has a good biotechnology base, but needs to strengthen this base to be competitive. In like manner, the future of Africa requires first- and third-world partnerships—among governments, scientists, private sectors, agricultural sectors, resource-poor farmers, and NGOs—in order to ensure the benefits of this technology for all.

NOTES

1. FAO, "Agricultural Biotechnology for Developing Countries." Results of an Electronic Forum, J. Ruane and M. Zimmermann, forum coordinators, FAO Research and Technology Paper 8, 2001.

2. FAO/GIEWS, "Food Supply Situation and Crops Prospects in Sub-Saharan Africa," *Africa Report* no. 3 (December 1998).

3. FAO, 1999, www.fao.org/NEWS/1999/991004 e.htm.

4. E.R. Orskov, *Reality in Rural Development Aid with Emphasis on Livestock* (Aberdeen, U.K.: Rowett Research Services Ltd, 1993), 11.

5. FAO, 2001.

6. J.A. Thomson, *Genes for Africa—Genetically Modified Crops in the Developing World* (Landsdowne, South Africa: UCT Press, 2002), 155.

7. G. Conway, *The Doubly Green Revolution: Food for All in the Twenty-first Century* (Ithaca, N.Y.: Penguin, 1997), 151–152.

8. Conway, 151–152.

9. Comprehensive Africa Agriculture Development Programme (CAADP), NEPAD, 2003.

10. C. James, *Global Review of Commercialized Transgenic Crops*, ISAAA Brief No. 24 (Ithaca, N.Y.: ISAAA, 2003).

11. Thomson, 160.

12. M. Qaim, *The Economic Effects of Genetically Modified Orphan Commodities: Projections for Sweet Potato in Kenya*, ISAAA Brief No. 13 (Ithaca, N.Y.: ISAAA, 1999).

13. Thomson, 162.

14. Thomson, 162.

15. M. Blakeney, "Agricultural Research and the Management of Intellectual Property," in *Managing Agricultural Biotechnology—Addressing Research Program Needs and Policy Implications*, ed. by Joel I. Cohen (The Hague, Netherlands: CABI/ISNAR, 2001), 228.

16. FAO, 2001.

17. M.V. Montagu and D. Oliveira, "Science and Society; The Ethical Responsibility" (presented at Towards Sustainable Agriculture for Developing Countries: Options from Life Sciences and Biotechnologies: European Group on Life Sciences–European Commission Research DG Brussels, January 30–31, 2003).

18. G. Acquaah, "Don't Trample Biotechnology; Proceed Carefully and Ethically," (2001), www.whybiotech.com/en/news/con780.asp?MID=17.

11

Autonomy, Humane Medicine, and Research Ethics: An East Asian Perspective

David Kum-Wah Chan

Do principles of bioethics differ between cultures? In the United States, bioethics has become an important subject in the last forty or so years, and autonomy has been enshrined as one of the key principles of medical ethics, alongside beneficence, nonmaleficence, and justice.[1] In recent years, the paternalistic model in the physician–patient relationship has been completely replaced by a model that stresses the patient's informed consent before treatment by the physician.

Since the concept of autonomy has for centuries played an important role in Western philosophy, as well as in the rise of liberal democracy, its role in medical ethics has been long overdue. But the concept of autonomy is not found in many non-Western societies, and has only arrived on the scene there as an imported Western idea. It evidently is not found in the philosophy of the ancient Chinese sage Confucius that has deep historical roots in East Asian countries. Confucian ethics articulates an authoritarian and hierarchical system both for the state and in the family, and the community's interests are valued above the individual's. It is thus difficult to envisage the principles that developed in Anglo-American bioethics as universalistic if such principles are to include autonomy. If an Asian bioethics based on ancient Confucian ideas could not incorporate a principle of autonomy, could it develop a non-Western alternative that could serve East Asian societies where modern medicine has gained common acceptance?

This chapter describes the basic values of medical practice according to ancient Chinese medical ethics, and examines whether an ethics based on the ideas of Confucius can sufficiently guide physicians in their relationship with patients. I suggest that in the arena of research, a recent case in Singapore illustrates how traditional medical ethics in East Asian societies is inadequate and needs to be supplemented by a principle of autonomy. Therefore, where research ethics is concerned, the principle of autonomy may well have validity across cultures, and autonomy will become an important value as biotechnology spreads to countries in East Asia.

ANCIENT CHINESE MEDICAL ETHICS

Confucian thought has influenced every aspect of Chinese life for more than two thousand years. This philosophy spread to neighboring countries in East Asia through Chinese imperial power, the spread of Chinese culture, and later by way of Chinese migrants who settled in Southeast Asia. Today, Confucian ideas continue to be studied and taught in Taiwan, South Korea, Hong Kong, and Singapore, countries that became known as the "Asian tigers" for their accelerated economic growth and development. Rapid modernization in these East Asian societies has included provision of modern health care modeled on Western medicine. There has, however, been great suspicion concerning Western ideas of democracy and individual autonomy.

Western-educated physicians in East Asia, like their counterparts in America or Britain, have some exposure to the subject of biomedical ethics. However, as in many places in the West, their training in ethical thinking is often inadequate. Unlike in the West, physicians in East Asia have also to integrate Western bioethics with Confucian values found in their own society. In addition, Confucian ethics has not been systematically applied to moral problems in medicine. Ancient Chinese medical ethics has been preserved in the form of codes and rules of conduct that were compiled by various Confucian medical scholars in earlier times. It is likely that aspects of Confucian ethics have surfaced in medical practice due to the cultural background of medical practitioners in East Asian societies rather than to the study of these historical texts.

Awareness in Western countries of an Asian bioethics has derived from infrequent attempts at cross-cultural comparisons that question the universalistic aspirations of Western bioethics.[2] The earliest available Chinese text on the duties of physicians to their patients was written by Sun Szu-Miao in the seventh century. Sun emphasized compassion and humaneness as the basic values of medical practice. Almost a thousand years later, a similar emphasis was repeated in Kung Hsin's maxim of 1556: "The good physician of the present day cherishes humaneness and righteousness. . . . His beneficence is equal to that of Providence."[3] The idea of humaneness is at the heart of Confucian ethics and is the defining virtue of the ideal Confucian man:

> It was widely accepted that a physician's saving his patients' lives and promoting their welfare was as respectable as a Confucian scholar's realizing his moral and political aspirations through ruling the states and bringing peace and pros-

perity to people. The well-known saying: "The achievement equals that of a good prime minister," frequently used by Chinese people to praise a successful physician, reflects this concept exactly. . . . The practice of medicine is the realization of humaneness.[4]

The virtue of humaneness is realized in the practice of medicine through an attitude of kindness and benevolence. The principles laid down in the ancient texts instruct physicians to diligently master the skills of their profession, to be selfless and devoted to their patients, to treat rich and poor equally, and to be modest, dignified, and respectful. Emphasis is placed on the value of life and the patient's welfare, not the physician's own glory and self-aggrandizement. However, there is no requirement for the physician to respect the wishes of the patient. Given that individual autonomy is not valued for its own sake in Confucian ethics, this is unsurprising. Firstly, there is a larger role for the family or community to make medical decisions for the individual. Secondly, the hierarchical system of authority places the educated physician above the common person.

Without concern for autonomy, physicians guided by humaneness will do their best to look after their patients in a paternalistic manner. Ancient Chinese medical ethics is characterized as beneficence-oriented in contrast to the autonomy-oriented medical ethics found in Western countries.[5] But paternalistic practices are not alien to Western medicine, as the Hippocratic oath at the foundation of Western medical ethics does not require physicians to respect the wishes of patients. Even today, the balance between beneficence and autonomy is debated in Western countries. Beneficence is thus a value that is recognized in more than one cultural context, and ethical physicians in both East Asian and Western societies are guided by beneficence in their conduct toward their patients.[6] The absence of a principle of autonomy in ancient Chinese medical ethics allows physicians to be more paternalistic than their Western counterparts, but this is culturally acceptable in traditional East Asian societies and not necessarily unethical.

PRINCIPLES OF RESEARCH ETHICS

Medical progress requires the development of new drugs and treatments. Research is necessary to discover how the body works, what illness does to the body, and how the patient responds to treatment. Since it is humans who would be treated, testing has to be done on humans to find out a treatment's effectiveness as well as side effects. The history of medicine is replete with examples of patient well-being and lives being sacrificed for the sake of science. What is clearly reprehensible about these examples is how human beings are treated as objects and used to serve the purpose of the researcher. Impetus for a code of research ethics for human experimentation came at the end of World War II, during the trial of German scientists who had experimented with a callous disregard for the suffering they inflicted on their human subjects. The Nuremberg Code of 1948 emphasized the centrality of voluntary consent. Its first article states: "The voluntary consent of the human subject is absolutely essential." Such research must be absolutely necessary (Article 2), and all unnecessary suffering and injury must be avoided (Article 4).

The main ethical problem for physicians who conduct experiments on patients is that they are violating the principle of nonmaleficence in deliberately risking harm to the patient, and the principle of beneficence in withholding treatment of patients in a control group. The Declaration of Helsinki, adopted by the World Medical Association in 1964, distinguishes between therapeutic research where subjects of research stand to benefit if the treatment on trial works, and nontherapeutic research where there is no such benefit for the test subject. Therapeutic research derives some moral justification from the principle of beneficence, whereas nontherapeutic research cannot be justified except on the ground that the subject voluntarily and knowingly assumed the risk. Thus, the informed consent requirement is more stringent in the latter case.[7]

Research on humans is now part and parcel of medical science in Western countries. The dual role of physician and researcher clearly changes the physician–patient relationship. Physicians would not have a moral basis for using (or allowing for use) their patients for research without an ethics of informed consent. A physician is justified in limiting the duties of beneficence and nonmaleficence only because patient autonomy is respected.

Medical research has spread beyond Western centers of research for a number of reasons. Scientific expertise is being tapped in non-Western countries. The development of treatments needs to take into account different responses from different test populations. And scientific validity may well depend on a large enough population of test subjects that cannot be found in Western countries alone. All of these factors are applicable to what is perhaps the most important medical research in human history: research in human genetics.

Non-Western countries embarking on medical research are included under the Declaration of Helsinki. They need to observe the requirement of informed consent for experimentation on human subjects. Can this requirement be adopted if individual autonomy is not traditionally valued in these countries? What needs to be done for research to be carried out ethically in East Asian countries where Confucian values still form the basis of traditional ethics and where autonomy is an imported moral concept? Answers to these questions are not yet fully available, but lessons can be learned from the experience of Singapore.

A CASE STUDY

Since the early 1990s, Singapore has embarked on a project of becoming a world biotechnology center. The country was already known for its high standard of health care. As a former British colony, its doctors are trained in Western medicine and its medical specialists are accredited to British professional institutions and colleges. With a well-educated population, the country's universities can be counted on to produce scientific researchers, many of whom undergo postgraduate training in the United States and Britain. In addition, the government built infrastructure and offered tax incentives to attract biotech companies and top scientists from Western countries. Moreover, the National Medical Ethics Committee has served to ensure high ethical standards for Singapore's medical professionals. In the year 2000, the Bioethics Advisory Committee was appointed by the government to propose legal and ethical guidelines for biotechnology research in Singapore, modeled on Western precedents. In its first term, the Committee produced a report after examining issues relating to both stem cell research and research in human genetics.

On January 19, 2003, the people of Singapore were shocked to read a story in their local newspaper regarding a possible breach of ethics in the conduct of a research project that was headed by the director of the National Neuroscience Institute (NNI). The details that came to light after a thorough investigation by an inquiry panel of independent experts are summarized here.[8]

The NNI had been set up in March 1998. Besides patient care services, it has a research arm with facilities for neuroscience research. In December 2000, British Professor Simon Shorvon was appointed director of NNI. Shorvon obtained a grant from the Singapore Biomedical Research Council (BMRC) for a research project he was conducting to identify genes that influenced susceptibility to Parkinson's disease (PD), epilepsy, and tardive dyskinesia, and also genes that influenced responsiveness to drug treatment. The method of research involved collecting blood samples from subjects with the neurological diseases (and from persons without these diseases as a control group), extracting DNA from the blood, and lab analysis of the DNA. In October 2001, Shorvon's Ph.D. student, Dr. Ramachandran, was appointed project manager to work under Shorvon's supervision.

The complaints that surfaced after November 2002 concerned the way in which patients with PD were identified, contacted, and tested. Shorvon needed blood samples from 750 PD patients by July 2003. By July 2002, only twelve volunteers had been recruited through their neurologists. Shorvon and Ramachandran obtained a list of PD patients directly from hospital records and without their neurologists' knowledge. They also devised a procedure to confirm that the patients indeed suffered from PD, a procedure that involved taking patients off their usual medication for at least twelve hours, observing their movement disorders, and then administering a "standard" dose of L-Dopa to observe their reactions. They had obtained neither approval from ethics committees nor informed consent from the PD patients for the procedure, while the patients' own neurologists were kept in the dark. Moreover, testing was performed in patients' homes or at NNI, without these patients being admitted to a clinical center for proper care and medical attention. Some patients had adverse reactions to the procedure due to omission of their normal medication or due to the administration of a different dose of L-Dopa during the testing.

The inquiry panel appointed by NNI concluded in its report of March 21, 2003, that the research project had been unethical in four areas: breach of confidentiality, lack of ethics committee approval, compromise of patient well-being and safety, and failure to obtain informed consent. It held Shorvon and Ramachandran clearly responsible for ethical misconduct.[9] In describing the procedure used in the project, the panel stated that PD patients had been treated for the purpose of research as "experimental subjects without any rights" (paragraph 73), in a manner "unacceptable in any civilized country" (paragraph 344).

ANALYSIS AND COMMENTARY

The inquiry panel in Singapore had done a superb job in a very short time to gather evidence on the case and document the ethical violations of the research carried out by Shorvon and Ramachandran. In presenting this case study, my purpose is to draw out some cross-cultural issues in research ethics. One possibility is that there is no such issue here: the case is one of a simple failure to follow universal guidelines for experimentation on human subjects. The Declaration of Helsinki had made patient consent obligatory for nontherapeutic research, and also mandated ethics committee approval for all research on human subjects. In Singapore, the National Medical Ethics Committee had also published guidelines that reflect those in the declaration. Shorvon is a British doctor who should have been aware of the ethics of human experimentation.

One issue that does arise in the case is whether the ethical lapses in the research project could have happened in the West. With respect to accessing patient confidential data through hospital records, Professor Burgunder, a coprincipal investigator of the project, told the panel that this would "never have been allowed in Switzerland" (paragraph 105). The list of hospital patients suffering from Parkinson's disease was obtained by Ramachandran from the pharmacies of two hospitals, after he stated that the research study was of "national interest." Furthermore, Shorvon had authorized the application for information, falsely giving an assurance that release of confidential patient information had ethics committee approval. The panel mentioned that the application had "obviously carried greater weight because he was Director of NNI" (paragraph 83).

With respect to the failure to obtain informed consent from PD patients who were used as test subjects, patients interviewed by the inquiry panel stated that they were under the mistaken impression that the testing was done with the approval of their treating neurologists (paragraph 266). None of them were aware of the risks associated with the tests performed on them, and none were given explanations as to the nature of the L-Dopa tests. It never occurred to them to question the authority of Ramachandran or the justification for having their medication modified. In fact, they had omitted their medication prior to testing after receiving a phone call request to do so from one of two persons working for Shorvon, neither of whom were qualified medical professionals.[10]

Both the violation of patient confidentiality and the failure to obtain informed consent reflect a breach of the principle of respect for autonomy. Shorvon had knowingly neglected his duty to patients and had acted unethically. But what made it easier for him to engage in ethical misconduct was the failure of a number of people to protect the autonomy of patients. Those in the hospitals who were impressed by Shorvon's status as director of NNI failed to protect confidential patient information. Those who worked for Shorvon failed to raise critical questions.[11] Most surprising of all was the failure of the patients or their family members to ask what the tests were about and whether there were risks to the patients. Yet the mere assumption that their own neurologists knew about the tests should not render informed consent unnecessary.

It is entirely possible that these failures to protect patient autonomy are due to the fact that the concept of individual autonomy is absent from East Asian values in general, and from ancient Chinese medical ethics in particular. Even though health care and the medical profession in Singapore have a Western orientation, it does not seem that the imported value of individual autonomy has been fully integrated into the nation's medical professionals' and patients' practices and attitudes.[12] Another aspect of Confucian ethics is respect for those in authority and for those with high educational qualifications. Thus, patients did not find it necessary to ask questions about what was being done to them. Those under Shorvon did not question his authority.

As described above, ancient Chinese medical ethics is beneficence-oriented. Physicians are expected to place patient well-being over the physician's own interests. Humaneness is a virtue that characterizes the life of every person who exemplifies Confucian ideals. Traditionally, education in China included training in Confucian ethics, and appointment to important posts was based on success in examinations. There was an assumed link between virtue and position. So another reason that the actions of those in authority are not questioned is the belief that such persons are motivated by benevolence. In Shorvon's case, this assumption proved to be completely mistaken.

In ethical lapses of enormous gravity, he and Ramachandran used a procedure that varied PD patients' medication in ways that endangered their well-being and safety. The inquiry panel described their "callous disregard for the PD Subjects' welfare" (paragraph 200). PD patients were instructed to omit their medication even though the resultant movement disabilities would cause them discomfort and make it difficult for them to get out of bed. Some of them were asked to travel to NNI even though they risked falling down. Omission of medication could bring about high fever, kidney failure, hyperthermia, and intense spasms. The increased dosage some patients were given during the testing could cause their tongues or limbs to become twisted. One patient (identified by patient folder number as PD10) experienced a fall in blood pressure from 140/100 to 100/70 within fifteen to twenty minutes of receiving a test dose that was four times his usual dosage (paragraph 202). Another patient (identified as PD 196) "suffered dyskinesia in both lower limbs, dystonia in his left hand and a marked systolic blood pressure fall of 50" (paragraph 202). The panel found that no assessment was made as to whether the test subjects could skip medication or travel to NNI, and no provision was made to admit subjects into clinics, leaving them at home without medical attention while they skipped their medication.

Such utter disregard for the test subjects' well-being and safety is a violation of ethical principles in both Western bioethics and ancient Chinese medical ethics. In the West, however, a failure to fulfill the duty of beneficence (or nonmaleficence) may not be quite so damaging because physicians are expected to respect patient autonomy by obtaining informed consent. The absence of a duty to respect patient choice in traditional medical ethics in East Asia leaves open the possibility that patients, who do not expect to be asked for informed consent, can be harmed by physicians who take advantage of their authority. Physicians such as Shorvon who come from Western countries may be tempted to do things that they would not do back home.[13] Where research, especially the nontherapeutic kind, is concerned, failure to ensure that test subjects are volunteers leaves them vulnerable to risks that are taken only to serve the interest of researchers. It seems clear that ancient Chinese medical ethics is not quite equipped to guide ethical research.

CHALLENGES AND ALTERNATIVE VIEWS

There are two ways in which my analysis of the Shorvon case may be challenged. First, it may be said that there is no ethical system that can guarantee that physicians act ethically. Shorvon and Ramachandran acted unethically because they were willing to disregard ethical rules. If they had been concerned about their test subjects' well-being, as required by the traditional ethical standards of East Asia, they would not have exposed them to danger, and would not even have accessed their records without the patients' own neurologists' knowledge. Adding a principle of respect for patient autonomy to ancient Chinese medical ethics would not alter the unethical conduct of researchers who do not care about ethical rules that require informed consent from their subjects.

But my point in discussing the Shorvon case specifically concerns research ethics. It is in the nature of research that the benefits accrue to others and not just, if at all, to the patient. It is also part and parcel of research that test subjects have to bear risks of harm, the existence of which the researchers aim to discover. Thus, physicians who are guided only by beneficence and non-maleficence toward their patients cannot possibly find it ethical to use them as test subjects. The only ethical justification for conducting research on patients is the fact that the patients have voluntarily agreed to participate in the research with full knowledge of the procedures and the dangers. In the absence of informed consent, research on human subjects cannot be ethical. Of course, Shorvon and Ramachandran are unethical by the standard of ancient Chinese medical ethics. But relying on that standard, no research on humans could be ethical. It is by appeal to the principle of respect for autonomy that the difference between ethical research and the kind of research carried out by Shorvon and Ramachandran can be articulated.

A second challenge to my analysis comes from those with the view that the ethics of the Shorvon case can be interpreted entirely in terms of Western medical ethics. The problem is not that the concept of patient autonomy has not been fully absorbed into the culture, but that there are different ideas about what informed consent means in Shorvon's research. Disputes about what counts as informed consent have often given rise to ethical problems in Western countries.[14] If this interpretation is right, then no cross-cultural issue is raised by this case.

Shorvon had attempted to defend himself before the inquiry panel by suggesting that the form used to obtain consent from PD subjects, which had been submitted for ethics committee approval, covered the procedure of L-Dopa testing, since it included the line: "You will also be asked for information on yourself and your condition" (paragraph 275). The panel was swift and devastating in demolishing this explanation: "No one can seriously suggest that the phrase 'information on yourself and your condition' covered the L-Dopa responsiveness testing, which involved omitting and varying medication." It is important to note that the PD subjects did not assume that they were consenting to something else, but that they had cooperated with the testing without having signed any consent form. None of them asserted a right to an explanation of the test procedure, purpose, and risks, nor were they initially aware that they were participating in research as test subjects.[15] Thus, different ideas about what constituted informed consent were not the issue here. Rather, it was the absence of any expectation on the part of PD subjects that the physician is duty-bound to obtain informed consent from his or her patient or subject before proceeding with treatment or research.

LESSONS TO BE LEARNED

The universality of American bioethics has been challenged by cultural pluralists, who perceive that "bioethics is dominated by the West and by the Western ethos of liberal individualism."[16] Ancient Chinese medical ethics provides an alternative model of the doctor–patient relationship that is beneficence-oriented. But this model does not seem suitable for an ethics of research. In research, physicians have to restrict the application of principles of beneficence and nonmaleficence toward their patients. The difference between ethical and unethical research is whether human test subjects have given informed consent. The principle of respect for patient autonomy may well have valid application across cultures wherever research is done.

The case study discussed here entitles us to say something stronger. In moving into biotechnological research, Singapore has also recognized informed consent as an essential ethical requirement. This did not prevent the unethical research that Shorvon engaged in. It is true that no set of ethical principles can prevent abuse by unethical individuals who disregard them. But those who worked with Shorvon and in the hospitals, and the patients themselves, did not value patient autonomy sufficiently. These parties did not expect patients to have a right to informed consent. To change this, it is not enough to use informed consent for safeguarding against unethical research. Respect for patient autonomy must become part of the entire culture of medicine in the country. For any country that joins the global enterprise of medical research, respect for autonomy becomes a universal ethical principle not only in research but in the practice of medicine in general.

NOTES

1. Tom L. Beauchamp and James F. Childress, *Principles of Biomedical Ethics*, 5th ed. (New York: Oxford University Press, 2001). This has been a highly influential textbook in which the four-principle approach to biomedical ethics is articulated.

2. Tao Lee, "Medical Ethics in Ancient China," in *Cross Cultural Perspectives in Medical Ethics: Readings*, ed. Robert M. Veatch (Boston: Jones & Bartlett, 1989); Daniel Fu-Chang Tsai, "Ancient Chinese Medical Ethics and the Four Principles of Biomedical Ethics," *Journal of Medical Ethics* 25 (1999): 315–321.

3. According to Tao Lee, "Medical Ethics in Ancient China," the complete maxim has been used as a motto in the baccalaureate service of the Peiping Union Medical College since 1939.

4. Tsai, 316–317.

5. Tsai, 320.

6. I take beneficence here to include nonmaleficence, since looking after the patient's well-being includes both benefiting them and not harming them.

7. The Declaration of Helsinki allows proxy consent on behalf of legally incompetent patients for the purpose of therapeutic research only.

8. The report of the Inquiry Panel has been published on the website of the National Neuroscience Institute, which can be found at www.nni.com.sg.

9. Ramachandran fled Singapore just as the ethical issues became public knowledge and refused to cooperate with the inquiry panel. The panel criticized Shorvon for changing his story during the inquiry, but he subsequently accepted the panel's report in a letter dated March 31, 2003.

10. Of the two persons, one had no medical qualifications and the other was a nurse. Both were employed for clerical and administrative work. In addition, a man who assisted Ramachandran in his testing of patients had previously worked as a security guard and had no prior experience in the medical profession.

11. Professor Burgunder claimed in his testimony before the panel that he was uncomfortable with the procedures and against accessing patient confidential information and bypassing the patients' own neurologists. But he did not bring his concerns to the attention of authorities. The panel felt that he had a share of responsibility for the ethical lapses of the project.

12. My claim here is supported by some of the data from a study that I collaborated on, and that has been published in D. Chan and L.G. Goh, "The Doctor–Patient Relationship: A Survey of Attitudes and Practices of Doctors in Singapore," *Bioethics* 14 (2000): 58–76.

13. When asked during the inquiry about his failure to obtain approval from the relevant ethics committees for accessing confidential medical records and testing patients by varying their L-Dopa medication, Shorvon claimed that his approach to obtaining ethics approval was "perfectly okay" by U.K. ethical standards. The panel however found this difficult to accept, as the Helsinki Declaration is a universal standard for research ethics (paragraph 176).

14. For instance, can a patient who is emotionally upset provide informed consent? And what is the status of unwritten wishes regarding the withholding of life-saving treatment?

15. They were only asked to sign a consent form (approved by an ethics committee) for the extraction of blood after the L-Dopa test procedure had been carried out (paragraph 67).

16. Segun Gbadegesin, "Bioethics and Cultural Diversity," in *A Companion to Bioethics,* ed. Helga Kuhse and Peter Singer (Oxford: Blackwell, 1998), 25.

12

Indigenous Knowledge, Patenting, and the Biotechnology Industry

Stella Gonzalez-Arnal

Within the knowledge economy, the most important economic asset is clearly knowledge. It has been argued that the ability to produce patentable knowledge in some key areas of the economy, particularly in those related to biotechnology, can change the fate of whole countries. Juan Enriquez, who endorses this view, claims that patents "are a good barometer . . . of creativity . . . tenacity. . . ability to articulate an idea . . . and capacity to build knowledge. Patents are a good window . . . (although not the only window) . . . on who might triumph . . . and who might lose . . . over the next two decades."[1] He also believes that to be able to obtain the maximum benefit from one's knowledge in an era of global commerce, there is a need to patent globally. Although different countries have different regulations, there are international agreements that regulate international commerce, and thus there is a tendency to homogenize the law to obtain global protection. Within this scheme, if wealth follows knowledge, less developed countries should have a good chance of improving their economies, as they have a fair amount of knowledge that could be useful for the biotechnological industry. But, in a less optimistic analysis, Diane Fowles makes us aware that there is a clear relationship between power, patenting, and knowledge: "Does the fact that only 1 percent of the patents go to Third World nationals mean that only 1 percent of the innovation on this planet is done in Asia, Africa, and Latin America? Or, might it reflect a bias in the system, an indication of who has written patent conventions and for whom?"[2]

This chapter analyzes the claim that current international agreements on intellectual property, particularly those on patenting, are inadequate for protecting traditional and indigenous knowledge. I also examine the assertion that, if these international agreements are globally enforced, indigenous and

139

traditional forms of knowledge and, consequently, their associated ways of life could be endangered. I argue for the protection of these indigenous communities and their ways of life by applying epistemological considerations. I also argue that the existence of a plurality of systems of knowledge actually benefits all peoples.

Here is an important caveat at the outset: Traditional and indigenous knowledge systems are quite diverse and there are different tools in intellectual property law (IPL) that can be useful in protecting some forms of knowledge within those societies. Nevertheless, I focus on some of the characteristics frequently linked to these alternative types of knowledge that make it especially difficult to cover some knowledge practices through current types of legislation.

INADEQUACIES IN CURRENT PATENTING

Reasons why patenting in its current form is inadequate for the protection of traditional knowledge include the following: it is not affordable; it is culturally biased towards Western epistemological models; and its application does not reflect the importance of maintaining a diversity of knowledge practices. Regarding the first reason, it is generally accepted that the costs of applying for patenting and, furthermore, the legal costs of defending one's ownership if the patented product is pirated, exclude some knowers and even entire communities from being able to claim ownership over the knowledge that they hold. Thus it is clear that one of the aims of a global system on intellectual property should be affordability.

As to the second and third reasons, the nature of the bias toward Western epistemological models can be understood if we consider how patents have been granted on knowledge that was in the public domain within traditional and indigenous communities, and that, according to Third World activists, is only novel within the context of the ignorance of the West.[3] Inventions have to be novel in order to qualify as patentable. Therefore inventions that are in the public domain are taken to be "prior art" and are excluded from patenting. And even though the World Intellectual Property Organization (WIPO) admits different ways of public disclosure, it still acknowledges that the only type of disclosure that their officials investigate is whether or not an invention appears in publications.[4] Traditional knowledge holders can thereby be disfavored since their knowledge is often not published. The result? Outsiders are granted patents on "inventions" that are already broadly known by a traditional or indigenous community. This has been possible because their knowledge was not "appropriately" documented and therefore not considered "prior art." This has prompted some critics to denounce the biopiracy of the West, plundering the intellectual commons of developing countries.[5]

Within Western epistemology, with very few exceptions, the most valuable form of knowledge is that whose content can be explicitly articulated in a propositional form. In contrast, practical, tacit, and embodied forms of knowledge, which by their very nature cannot be articulated (or only partially so), are ignored. Furthermore, knowledge and expertise is mostly attributed to those who can make their knowledge explicit by means of articulation. Those who engage in a practice but who are not able to reduce it to a propositional form are generally considered to be lesser epistemic agents.

Along with favoring a reductive form of codification, Western epistemology is also biased towards theorizing. Traditional knowledge is often transmitted neither as theories nor as a body of knowledge in which knowledge practices are described propositionally, but rather as the correct way of doing something. Within Western epistemology, this characteristic signals an allegedly inferior type of knowledge. Finally, in our Western tradition, the favored way of passing knowledge on has been to publish it, even though we have been warned, as far back as Plato, that the written word is an inadequate medium for apprehending and transmitting knowledge and knowledge practices. These epistemological preferences permeate the legal framework that determines what is in the public domain, and thus what counts as prior knowledge. If intellectual property is attributed on these bases, and if this system is globally imposed, this will spread the misplaced assumption that the only knowledge that counts is that which is articulable and published.

Traditional knowers are also disadvantaged by another assumption: If practical and tacit knowledge is not viewed as knowledge (as within the foregoing narrow characterization of knowledge), then it does not count as intellectual labor. It can therefore be conveniently ignored when property claims are made over particular innovations or inventions that have been developed on the bases of traditional or indigenous knowledge. This has been the case with certain products developed by scientists in the biotechnology industry. Rural Advancement Foundation International has claimed that "no matter how stunning their technological achievements or costly their research, genetic engineers are literally building on the accumulated innovations and success of generations of anonymous farmers (as well as formal sector breeders)."[6] This can be illustrated by the following example offered by environmentalist and activist Vandana Shiva. For centuries Indian farmers have cultivated different varieties of basmati rice, which is one of their fastest growing export items. In 1997 RiceTec, a company based in Texas, was granted a patent on basmati rice lines and grain. The patent "covers the genetics lines of the basmati and includes genes from the varieties developed by farmers. It thus automatically covers farmer's varieties and if enforced, farmers will not be able to grow these varieties developed by them and their forefathers without getting permission for and paying royalty to RiceTec.

. . . As the RiceTec line is essentially derived from basmati, it cannot be claimed as novel and therefore should not be patentable."[7]

There is a clear bias toward Western science when establishing what can be patented, as even knowledge that has been articulated by traditional communities and that is part of their theoretical bodies of knowledge can be denied the benefits of patenting. WIPO claims that if an Ayuverdic physician who developed a new formulation to treat "wind disorders" "described his innovation in terms of the appropriate Ayuverdic diagnosis and aetiology, the innovation could not reasonably be considered an invention within the reason of the modern patent system."[8] WIPO itself acknowledges that the real value of traditional knowledge is "often overlooked within the modern reductionist approach to science. Unless information is developed under aseptic clinical conditions by scientific methods, it is sometimes viewed as 'inferior.'[9] Within this context, a further form of protection afforded by IPL, that of defensive publication, could be denied to traditional knowers as it would have to be made public within accepted knowledge discourses in order to deter others from patenting their knowledge. For instance, Indian farmers have known of the properties of the seeds of the neem tree as a pesticide for centuries. Yet, if they wanted their knowledge protected via defensive publication, then "Indian farmers would have had to isolate and name the active ingredients of neem, then publish the details to prevent companies from applying for patents."[10] The rationale of IPLs is that of promoting and rewarding intellectual labor, but in their current form, only certain types of intellectual labor are recognized. As it stands, the system is often incapable of rewarding innovations within traditional systems of knowledge. Within this context, it is hardly surprising that developing countries are encouraged to educate their citizens in Western style science and technology if they want to improve the economy of their countries. This seems to be sound strategic advice. However, as I later argue, if traditional knowledge is not also supported, one of the most important sources of innovation will be lost.

EPISTEMIC PRACTICES AND THE SCIENTIFIC VIEW OF THE WORLD

The fact that tacit and practical knowledge is considered inferior to explicit and codified knowledge, and that science is the measure of all other epistemic practices, is linked to the idea that science represents the world as it is and that any rational being, given the correct information, ought to agree with this. In Western culture, there are very strong links between epistemic reliability, rationality, and a scientific view of the world, and those who do not comply are seen as lesser epistemic agents. This has been judged as a form of imperialism that could be further encouraged if our current system of IPL is globally applied without introducing enough diversity into its framework to allow for cultural variation.

This link between the scientific view of the world and that of epistemically reliable character current in our culture is brought out clearly by a classic example taken from anthropology. Philosopher Peter Winch challenges Evans-Prichard's interpretation of the epistemic character of the Azande. The Azande believe in witchcraft and assume that there are causal relations of a kind that our scientific tradition denies. The issue here regarding their beliefs is not so much whether they are justified in having them, but whether it is rational to persist in having them when science challenges them, as science is taken to reflect the world as it is, and in this regard it is superior to alternative systems of knowledge. In this context, Evans-Pritchard claims that "scientific notions are those which *accord with objective reality* both with regard to the validity of their premises and to the inferences drawn from their propositions [emphasis added]."[11] This view is challenged by Winch, as he claims that what constitutes "objective reality" is always theory-bound and, therefore, it is not possible to judge "from the outside" the agreement between the world and science, or the disagreement between reality and the Azande's worldview. Furthermore, Evans-Pritchard questioned the epistemic character of the Azande when he noted that the Azande would incur contradictions regarding their beliefs in witchcraft if they press their thought to its logical conclusions, but that they do not do this is due to their lack of interest in approaching the issue theoretically. Winch counters that these characteristics of the Azande's view of the world are only recognizable from our way of thinking, as "Zande notions of rationality do not constitute a theoretical system in terms of which Azande try to gain quasi-scientific understanding of the world."[12] Their view of the world, as is ours, is grounded on the way in which they relate to the world and on their practices (which are different from ours), all of which allow them to make sense of their environment and to manage their affairs. Their view of the world, therefore, is constrained by different rules, by a different sense of what constitutes an appropriate explanation or what constitutes rational behavior, given the circumstances.

It is important to remind ourselves of this classic discussion because many of the assumptions that underlie Evans-Pritchard's analysis are still current. For example, WIPO has organized several fact-finding missions all over the world to see if the specific needs of traditional and indigenous knowers could be fulfilled by our IPL system. In its report, indigenous and traditional knowledge was characterized as nonsystematic. This document was open to consultation and includes a response to this claim about traditional knowledge (referred to as TK):

> There is an extensive scientific literature attesting to the systematic, if locally specific, nature of traditional biodiversity knowledge. It is precisely these systems of

plant use, culture and knowledge that local communities seek to preserve through IP (formal or informal). *To deny their systematic nature already places TK at a disadvantage in terms of developing and applying appropriate IP tools.*[13]

What clearly emerges is that a certain epistemic model is being used and imposed as the paradigm of knowledge, that is, it is deemed necessary to cite scientific studies to legitimize indigenous knowledge.

It is also clear that there is an imbalance of power in the process of instituting and enforcing IPL globally. Those who are stronger are the ones imposing their own paradigm of what constitutes a reliable knower and justifiable knowledge. A global IPL system should reflect the fact that all knowledge, including science, is situated knowledge.[14] This would make us more aware of the fact that different systems of knowledge are neither reducible to science, nor translatable into one another.

THE EPISTEMIC VALUE OF TRADITIONAL KNOWLEDGE

There have been steps taken by the international community to recognize the epistemic value of traditional knowledge. The Convention on Biological Diversity (CBD), for instance, recognizes the role of knowledge held by local and indigenous communities in the conservation of biological diversity, and signatory countries have committed themselves to preserving this knowledge and to sharing any benefits obtained from it. Establishing mechanisms to facilitate the sharing of benefits obtained from the exploitation of indigenous knowledge is, obviously, very important, but it has been pointed out that even when an agreement is reached with traditional knowers for the exploitation of resources disclosed by them, their knowledge practices are not recognized as such.[15] Unless the epistemic value of these types of knowledge is accepted, we will not be able to develop appropriate legal structures in order to protect this knowledge and reward it properly.[16] This is why it is important to acknowledge the value of the *labor* of traditional and indigenous communities, as this is a step towards its acceptance as intellectual labor.

Another cultural bias in the IPL system is the idea of authorship, which lies at the basis of IPL. The idea is that innovation is due to individuals, that it is possible to point out not only who the innovator is but when the innovation has taken place. James Boyle asserts that this concept of intellectual creation originated in the eighteenth century when the element in craftsmanship in writing began to decline and the idea that inspiration, which emanates from the author, began to be important. Those with original genius could take materials from common sources and make them more valuable by mixing them with their labor. The persistence of the myth of the author damages the interest of traditional knowers because it imposes a way of judging what counts as innovation that is inappropriate in its application to many of their knowledge practices. Furthermore, as Boyle points out, if a dispute should arise between scientists of a chemical company and traditional farmers over the patentability of some agricultural product, "the farmers are everything authors should not be—their contribution comes from a community rather than an individual, from tradition rather than innovation, from evolution rather than transformation. Guess who gets the intellectual property right?"[17]Traditional knowledge is often characterized as communal and traditional knowers seem well aware of the social nature of their knowledge. In these communities it is not uncommon to value tradition over innovation (although this does not imply that traditional systems of knowledge remain unchanged). Furthermore, within such communities it is often not accepted that only some knowers should benefit from communal knowledge or should have a monopoly over its application. Legal theorist Robert Burrell claims that in China, for instance, for historical and ideological reasons, tradition is valued and the social nature of knowledge is underlined. Therefore, Chinese do not generally think that IPLs enforce important interests. "As a result, there is no public acceptance of the need for such laws and they do *not see activities which infringe them as morally wrong* [emphasis mine]."[18] Accordingly, some traditional knowers from West Africa, collaborating with WIPO, claimed that IPL officials should be "made to recognize the fundamental differences between Western systems and African customs and culture. . . . Asking local people to pay for the use of knowledge which they believe to be theirs would, it was explained, go against the traditional concept of ownership and stifle creativity."[19] Furthermore, some traditional knowers believe that it is inappropriate for anyone to benefit economically from some types of knowledge, such as medicine; they also challenge the idea that they own the knowledge they hold, as they regard themselves as custodians of this knowledge. They do not believe that their knowledge belongs to them, but rather to their community and to future generations. Such beliefs no doubt challenge some of the tenets in our current intellectual property system since claiming ownership over knowledge that has a social origin seems rather unfair. Furthermore, it can be argued that economical rewards are not an appropriate way to induce people to innovate and to reward innovation.

HOW DO WE REWARD THE
INTELLECTUAL LABOR OF COMMUNITIES?

The ways in which traditional and indigenous communities are pressured to organize themselves and transform their own ways of regulating knowledge practices in order to be able to benefit from IPL are no doubt questionable. The collective nature of traditional knowledge makes it difficult to determine who is the holder of knowledge as it might be present in different communities and among many of their members. Even when some piece of useful knowledge has been identified and is exploitable, and all parties are willing to reach an agreement regarding the sharing of the benefits, it is difficult to establish what would be the best way of doing this. A common complaint from traditional knowers is that in order to be able either to protect their knowledge or to benefit from it they are forced to become corporations, which is highly inappropriate in many cases as this undermines the traditional organization of communities and their values. Shiva claims:

> the transformation towards the corporate identity assumes the embracing of certain values associated with the dominant capitalist paradigm of development. Those values being profit maximisation, market competitiveness, efficiency optimisation, product and/or output focus, market driven and investor managed. The characteristics of corporations such as: limited accountability, lack of wide consultation, unequal delegation of power and control, hierarchical management and inequitable distribution of resources within the entity, represent the antithesis of the community as we in India understand them to exist. Globally, the community identity is being marginalized and subsumed by the corporate state structures of the dominant western paradigm. However, communities do have legal personality in the common law practices of indigenous cultures and indigenous jurisprudence.[20]

Two different issues are underlined. The first is practical. Given that traditional knowledge does not fit the model of knowledge supported by IPL, what is the best way of utilizing existing schema so as to reward the intellectual labor of communities? As we have seen, acquiring a corporate identity appears to be an inappropriate solution for some communities. There are cases in which several communities might possess a particular type of knowledge, or a community might consider that a particular piece of knowledge belongs to future generations as well as to those presently living. While some authors have acknowledged the difficulties, they have also pointed out that it is not an impossible task.[21] The fact that it is not possible to reward a *single* individual or community for some knowledge does not mean that the economic benefits obtained by it should not be shared. For instance, many indigenous communities in Paraguay and Brazil have used a natural sweetener called "Stevia rebaudiana," which has industrial applications. Darrell Posey and Graham Dutfield claim that companies that benefit from it should be asked to contribute a portion of their benefits in order to invest this in the conservation and development of this part of the world.[22]

The second issue is that traditional ways of regulating the production and exchange of knowledge have a normative value that should have legal recognition. In some traditional societies the ownership of knowledge is recognized and carefully regulated. Applying IPL would erode these traditional ways of regulating knowledge production and distribution. As some traditional knowledge holders have claimed, "This could affect how traditional healers speak to each other, their children, how they transmit their knowledge and teach their students."[23] A powerful argument against our current system of international IPL is that it does not encompass all the different systems of ownership that already regulate the exchange of knowledge.[24] This means that we need to protect a plurality of systems. There are already examples in certain countries that have already developed *sui generis* systems to protect some of their assets, particularly those related to agriculture. Some authors have postulated the need for a mixed system of protection in which communities enforce their own custom-based intellectual property systems and benefit from others that are global.

Certain alternatives have been suggested such as the development of traditional resource rights (TRR) and of collective intellectual rights (CIRs). The Global Coalition for Bio-Cultural Diversity has proposed the development of TRR. This initiative "can accommodate a wide range of relevant international agreements as a basis for a *sui generis* system of protection for indigenous peoples and their resources—that is a system that is unique and does not belong to an existing category of IPR."[25] According to supporters of TRR, these rights "go beyond other *sui generis* models in that they seek not only to protect knowledge relating to biological resources but also to assert the right of peoples to self-determination and the right to safeguard "culture" in its broadest sense."[26]

CIRs have been supported by the Research Foundation for Science, Technology and Ecology (India) and the Third World Network.[27] These intellectual rights challenge some of the values that underlie IPR. For example, they acknowledge the social nature of knowledge. They also claim that rights should be granted in perpetuity, since the knowledge held by communities also belongs to future generations. CIRs include mechanisms that would allow traditional communities to more easily prove that they are the legitimate holders of the knowledge. For instance, "if evidentiary proof of the knowledge is required, any declaration by the community in a manner and form accepted by the cultural practices of the community shall be sufficient evidence for its existence."[28] Supporters of CIRs also argue that any system regulating intellectual property should allow knowledge holders to either economically exploit their knowledge or else exchange it freely. And any economic benefits obtained by this exploitation should belong to the whole community. Such benefits ought to be used for "the protection, development, strengthening and maintenance of the community and its knowledge and resources."[29]

One of the core arguments against adopting our current IPL system on a global scale is that there is a lack of democratic representation in the international institutions in which IPLs are developed. It is argued that deciding to apply one system or another to protect knowledge is an issue that needs to be settled through negotiation with the communities involved.

THE NEED FOR PLURALISM

As I have said, international forums have already established the need for pluralism. They have recommended that negotiations with communities should take place to establish a fair exchange. Some of the difficulties that these negotiations entail could be overcome by increasing the representation of different communities on the international bodies, by accepting the value of different epistemic systems, and by introducing legislation that reflects and promotes multiplicity. However, there is one problem these means cannot solve. Engaging in this type of exchange can threaten the ways of life of many traditional societies. If their way of life disappears, their knowledge practices will be lost. There are communities that possess knowledge that is valuable for industry. Yet these same communities engage in economies whereby reciprocity, instead of accumulation and profit, is the rule. Such communities encourage their people to share their goods, creating an obligation between giver and receiver. This is the basis of their society. The idea that knowledge can be privately owned and exchanged for profit does not enter into their way of life. And incorporating such an idea radically changes their cultural system and erodes their traditional knowledge practices.[30] Admittedly, some communities have been able to incorporate, to some degree, this idea into their "economies of gift" by putting forward collective economic projects.[31] Nonetheless, many of these projects have resulted in failure.[32] Furthermore, even when the social group does happen to function within a market economy, there are certain types of knowledge that simply cannot enter into economic exchanges,[33] and the nature of this knowledge could change if certain types of economic systems are imposed.[34]

If, within the knowledge economy, it is possible to benefit from traditional and indigenous ways of knowledge, a global system for patenting knowledge must allow those communities to benefit from their knowledge and to protect that knowledge from being exploited without their consent. This global system should also acknowledge the values and epistemic biases within the current IPL system. It should allow for a diversity of normative practices regarding the regulation of knowledge, as these are a central part of how knowledge is produced and transmitted and lie at the core of epistemic practices.

A plurality of epistemic communities and practices brings about innovations that can be economically valuable. It also allows us to create better, more objective science. Theorists such as Sandra Harding and Helen

Longino have claimed that pluralism is necessary for objectivity. They have argued that theories produced by marginalized epistemic communities can indeed offer us original and positive alternatives that challenge our mainstream traditions. Harding, in particular, has shown how those at the epistemic periphery are more capable of discovering the biases that remain undetected in all accepted theories, allowing us to critically reconsider theories that have been widely accepted. In this way the epistemic interaction of a multiplicity of communities of knowers definitely benefits mainstream science. It allows those at the center to see the biases in their own theories whilst also providing alternatives. This allows us to encounter aspects of the world that would otherwise remain unseen. Thus, a more encompassing epistemology as the basis of intellectual property rights would enable us to protect and reward traditional and indigenous knowledge that would at the same time be of benefit to us all.

NOTES

1. J. Enriquez, *As the Future Catches You: How Genomics and other Forces are Changing Your Life, Work, Health and Wealth,* (New York: Crown Business, 2001), 138.

2. C. Fowler, "Biotechnology, Patents and the Third World," in *Biopolitics: A Feminist and Ecological Reader on Biotechnology,* ed. V. Shiva and I. Moser (London: Zed Books, 1995), 220.

3. V. Shiva, *Patents: Myths and Realities* (New Delhi: Penguin Books, 2001), chap. 4 and G. Dutfield, *Intellectual Property Rights, Trade and Biodiversity* (London: Earthscan, International Union for Conservation of Nature and Natural Resources, 2000), 64–69, show how some patents on inventions derived from traditional and indigenous knowledge have been challenged.

4. *Intellectual Property Needs and Expectations of Traditional Knowledge Holders: WIPO Report on Fact-Finding Missions on Intellectual Property and Traditional Knowledge 1998–9* (Geneva 2001): 36, www.wipo.int/globalissues/tk/report/final/pdf/part1.pdf.uspto.gov.

5. V. Shiva, *Biopiracy: The Plunder of Nature and Knowledge* (Boston: South End Press, 1997).

6. T. Smith, "Biotechnology and Global Justice," *Journal of Agricultural and Environmental Ethics* 11 (1999): 233.

7. Shiva, 2001, 57.

8. WIPO, 109.

9. WIPO, 214–5.

10. D. A. Posey and G. Dutfield, *Beyond Intellectual Property: Toward Traditional Resource Rights for Indigenous Peoples and Local Communities* (Ottawa: International Development Research Centre, 1996), 81.

11. P. Winch, "Understanding a Primitive Society," in *Rationality,* ed. B. R. Wilson (Oxford: Basil Blackwell, 1986), 80.

12. Winch, 93.

13. WIPO, 26.

14. On the issue of situated knowledge see D. Haraway, "Situated Knowledges: The Science Question in Feminism and the Privilege of Partial Perspective," *Feminist Studies* 14, no. 3 (1988): 575–599.

15. A. Zerda-Sarmiento and C. Forero-Pineda, "Intellectual Property Rights over Ethnic Communities' Knowledge," *International Social Science Journal* 54 (March 2002): 111.

16. The CBD is a step forward in this direction. G. Dutfield (2000) cites articles 8(j) and 18.4 of the CBD to support the claims that holders of traditional knowledge have knowledge and technologies that are relevant for the conservation of biodiversity and that "there is no justification for assuming (as many tend to do) that such technologies have a lower status than other technologies relevant to the convention," 33.

17. J. Boyle, *Shamans, Software and Spleens: Law and the Construction of the Information Society* (Cambridge, Mass.: Harvard University Press, 1996), 126.

18. R. Burrell, "A Case Study in Cultural Imperialism: The Imposition of Copyright in China by the West," in *Intellectual Property and Ethics*, ed. L. Bentley and S. Maniatis (London: Sweet and Maxwell, 1998), 207.

19. WIPO, 148.

20. V. Shiva, A.H. Jafri, G. Bedi, and R. Holla-Bhar, *The Enclosure and Recovery of the Commons* (Indu Prakash Singh: Research Foundation for Science, Technology and Ecology, 1997), 88.

21. See Zerda-Sarmiento and Forero-Pineda, and Posey and Dutfield.

22. Posey and Dutfield, 40.

23. WIPO, 102.

24. USPTO commented "Is it possible or even desirable, to establish a comprehensive, uniform set of rules at the international level to govern the use of traditional knowledge and folklore? At the very least, we wonder whether it is advisable to take such activity before individual countries have, in conjunction with the communities within their own borders, established their own regimes for protection within their own territories and gained experience in the application of that protection and its effect in the communities involved. We believe that WIPO parties should consider these issues carefully" and "Moreover, as the Draft Report indicates, there are so many different expectations, goals and native systems, for approaching ownership and the transgression of ownership that a useful, enforceable global system would be virtually impossible to create. Indeed, a 'one size fits all' approach might be interpreted as demonstrating a lack of respect for local customs and traditions." Quoted in WIPO, 226.

25. Posey and Dutfield, 3.

26. Posey and Dutfield, 95.

27. See V. Shiva et al. and Third World Network, "The Main Elements of a Community Rights Act," 1999, www.twnside.org.sg/title/comm-cn.htm.

28. Third World Network.

29. Third World Network.

30. Zerda-Sarmiento and Forero-Pineda, 104.

31. Zerda-Sarmiento and Forero-Pineda.

32. Zerda-Sarmiento and Forero-Pineda, 105.

33. Many traditional knowledge holders feel that they should not put financial value to their knowledge, particularly if it is for health-related purposes; see WIPO, 147.

34. The Future Harvest Centre points out that even when a system of collective ownership is applied to certain types of knowledge, in some societies, some problems might arise. For instance, in some agrarian societies women are responsible for community germplasm conservation. If their crops are given economic value, they will lose control over decision making as the men in their communities will take control; see WIPO, 213.

III

SPECIFIC GLOBAL CHALLENGES

Now that we've examined specific cultural perspectives, issues, and emphases, what are the challenges all of this presents within the global community? The first challenge, as seen in part II, is to understand and be more sensitive to the different cultural contexts concerning various applications of biotechnologies. The next challenge is to ask whether a shared vision of biotechnology is possible and feasible. How can we ensure equity when it comes to benefit-sharing? We thus come full circle. To address this need for equity, this section focuses upon three prominent global challenges: fair patenting, fair media discussion, and examining biotechnologies within the current context of the ongoing threat of bioterrorism.

Is it possible to strike a morally justified balance between patent rights and access rights? Dianne Nicol gets to the heart of this crucial question by examining stipulations and implications in recent international negotiations that seek to achieve this balance. Without this balance, biotechnology's promise of better drugs, diagnostics, and therapies is one-sided, clearly favoring those countries in the developed world that hold patenting rights. Surely, within these countries there is the implicit assumption that there is a right to obtain patents in order to provide further marketing incentives. Yet this comes into direct conflict with developing countries' right to have access to quality medicines and therapies, since products that are under patent are simply unaffordable for poorer countries.

Nicol also points out that allowing domestic manufacture of cheaper, generic drugs is not the solution since poorer countries lack the capacity to produce these drugs. Nicol provides the reader with an excellent introduction into the complex world of patenting. Moreover, she thoughtfully examines international efforts, such as the World Trade Organization's Agreement on

Trade-related Aspects of Intellectual Property (known as TRIPS), to restore some level of equity and to sustain an ethically legitimate balance between patent rights and access rights. In doing so, she astutely brings to our attention the most crucial concerns as well as practical issues in acquiring this balance.

The media plays an undeniably powerful role in shaping the contours of public perception and discussion when it comes to advances in biotechnology. Margaret Coffey examines the roles and responsibilities of media in this light. She points out that much media discussion remains detached from the cultural contexts of specific biotechnological applications. To illustrate, she reconsiders aspects of the Molly Nash case (described in detail earlier by Jeffrey P. Kahn and Anna C. Mastroianni) and shows how the media portrayal of the case was shortsighted in not considering all-important cultural perspectives and values that are not necessarily shared by other cultures.

She urges media workers as well as all professionals reporting on biotechnologies to be more sensitive to unquestioned assumptions behind the discussions. In her valuable discussion of "cultural engagement and media responsibility," she argues that fair media reporting requires a sensitivity and understanding of the diverse traditions and beliefs of other cultures as well as that of the reporters. Cross-cultural sensitivity and understanding is imperative in media reporting of biotechnologies. After all, as Coffey convincingly reminds us, biotechnologies are an inherently global enterprise in our global community.

Our last essay addresses what is perhaps the most distressing abuse of biotechnology in our world today—bioterrorism. This is especially disturbing since terrorism has no technological solution. Katharine R. Meacham and Jo Ann T. Croom prompts us of the razor's edge in that perennial adage, "knowledge is power." Thus biotechnological knowledge can be used to kill and destroy as well as to heal and give life. For the authors, biotechnology is "the twenty-first-century Trickster." The same knowledge that can be used to prevent disease and alleviate illness and disabilities can also be used to design bioterrorist agents.

Meacham and Croom offer an excellent and incisive account of what bioterrorism involves, along with current studies about both risks of and responses to bioterrorism. They also discuss realistic and fair preparations for what may be the inevitable. Applying themes from Camus's *The Plague*, they wisely argue that any response and any preparation can only be moral if we are willing to recognize and embrace each other across all religious, national, and cultural lines. Just as a morally sound view of biotechnology requires a shared vision, so our survival as a global community thus rests upon recognizing our common ground, and that, as Camus writes, "there are more things to admire in [human beings] than to despise." Their farsighted counsel is thus a fitting way to conclude our volume. Acting upon their advice, however, is a necessary beginning.

13

Cross-Cultural Issues in Balancing Patent Rights and Consumer Access to Biotechnological and Pharmaceutical Inventions

Dianne Nicol

Patents give patent holders and their licensees a period of market exclusivity, which enables them to charge monopoly prices on the sale of their products for the life of the patent, usually twenty years. Patent grants are generally justified on the utilitarian ground that they provide the necessary incentive to innovate and that innovation is good for society. The line of argument is that if inventors can recover the costs put into research and development and earn rewards through product pricing and license fees they will be encouraged to take their inventions through to commercial production and go on and create further inventions. The other side of the coin is that consumers are faced with having to pay high prices for patented products. The question is how to balance the public interest in encouraging innovation with other public interests. In the area of public health, for example, how should the public interest in equitable access to new treatments be balanced against the incentive to innovate that is provided by the patent system?

In the developed world, the balance swings strongly in favor of encouraging innovation, particularly with respect to drug development. Pharmaceutical patents are seen by the industry as being crucial to its economic viability. The reasons that are generally given for this are that research costs are high, the procedures for testing safety and efficacy are onerous and lengthy, and many potential products never make it to market. Biotechnology research and development may be even more risky because of high costs and the time lag between discovery and commercialization. In order to recover costs and to provide sufficient return on investment, products must be priced well in excess of production costs and must be protected from competition from cheaper generic copies. The patent system provides this protection. Without it (so the argument goes) the industry will discontinue its

155

innovative research and development. However, as a direct consequence, only those people who have sufficient wealth or who have access to subsidized health care are likely to have access to patented products. If we accept that in general the patent system has achieved the right balance between access and incentive to innovate (a conclusion that in itself is open to question), there may, nevertheless, be exceptional circumstances where the public interest favors adjusting that balance. This chapter explores the critical issue of how the balance might be adjusted in circumstances of public health crises in developing countries.

PATENTS AND ACCESS TO MEDICINES IN THE DEVELOPING WORLD

The World Trade Organization (WTO) sets the parameters for the intellectual property laws in all trading nations through its agreement on Trade-related Aspects of Intellectual Property (TRIPS). Article 27 of TRIPS requires that patents be made available for all inventions in all areas of technology, thereby mandating patents for pharmaceutical inventions. Members of the WTO discussed these issues at its Ministerial Conference in Doha on November 9–14, 2001, as a result of which the Declaration on the TRIPS Agreement and Public Health was promulgated.

In many countries, particularly sub-Saharan Africa, diseases such as HIV/AIDS, malaria, and tuberculosis are at epidemic levels. The seriousness of the problem is expressed succinctly in the first paragraph of the declaration:

> We recognize the gravity of the public health problems afflicting many developing and least-developed countries, especially those resulting from HIV/AIDS, tuberculosis, malaria and other epidemics.

Biotechnology is playing an important role in identification, diagnosis, and treatment of such diseases, and medicines are being developed to alleviate some of the suffering caused by them. However, the cost is prohibitive to the majority of the world's poor.[1] The urgent need to find some means of supplying pharmaceuticals at affordable prices is well recognized. A number of the companies that own patents to the antiretrovirals used in the treatment of HIV/AIDS have agreed to supply them at significantly reduced costs. However, while such gestures are to be applauded, they can only be partial solutions. The statistics are chilling. It has been reported that per capita health expenditures in low-income developing countries average $23 per year, yet the cost of the cheapest antiretrovirals is still $200 per year.[2] Affirmative steps need to be taken to ensure the delivery of affordable medicines to citizens of developing and least-developed countries, at least insofar as they relate to HIV/AIDS and other like epidemics, even if this means chang-ing the access–innovation balance.

ADJUSTING THE BALANCE: THE ROLE OF
COMPULSORY LICENSING

One option that is recognized as being likely to provide a satisfactory out-come is compulsory licensing.[3] A compulsory license is a court or adminis-trative order granting a license to work the invention without the authoriza-tion of the patent holder. Article 31 of TRIPS allows for use without authorization subject to certain limitations. The first restriction is that, prior to use without authorization, the proposed user must have made efforts to obtain authorization from the patent holder on reasonable commercial terms and conditions and such efforts must not have been successful within a rea-sonable period of time. This requirement may be waived in some situations including national emergencies and other circumstances of extreme urgency. However, there are still a number of limitations, perhaps the two most sig-nificant of which are that the intended use must be *predominantly for the supply of the domestic market* (Article 31[f]) and that reasonable remunera-tion must be paid to the patent holder (Article 31[h]). The relevance of these provisions is discussed later in this chapter.

The question addressed at Doha was whether the provisions in Article 31 allow for compulsory licensing of pharmaceutical patents to enable domes-tic manufacture of generic copies. For example, could the national emer-gency provisions be relied on to produce anti-HIV/AIDS drugs? These ques-tions were answered affirmatively in the declaration. In summary, the declaration confirms that compulsory licensing can be used as a means for protecting public health.

The importance of the declaration cannot be understated. For many, TRIPS was perceived solely as a means for developed countries to protect their corporate interests from piracy that is believed to be prevalent in de-veloping countries. It seemingly offers little in return to developing coun-tries in terms of protection from the full rigors of Western-style intellectual property laws, which focus predominantly on economic considerations. The declaration clearly indicates a fundamental shift in how TRIPS may be perceived. It shows that certain safeguards are embedded in the agreement. It can be used as a shield to enable measures to be taken to protect public health that are both consistent with the agreement and that at the same time do impinge on what would generally be considered to be traditional intel-lectual property rights.

The question that must be asked is, How effective are the Doha provisions likely to be in improving access to biotechnological and pharmaceutical inventions? There are few examples of compulsory licensing activity since Doha. Useful information about this activity can be found on the website of the Consumer Project on Technology, or cptech, at www.cptech.org. One example is Zimbabwe, where an emergency was declared in June 2001 for six months to enable manufacture or importation of antiretrovirals to be used in the treatment of HIV/AIDS. In South Korea, an application was made in January 2002 for a compulsory license to manufacture *imatinib mseylate* (trade named Gleevec or Glivec), a pharmaceutical composition for the treatment of cancerous tumors, the patent for which is owned by Novartis. A number of other countries are putting the necessary legal and administrative arrangements in place to allow for compulsory licensing.

There are various reasons why compulsory licensing activity has been slow to date. In some countries, compulsory licenses are not required at present because the necessary pharmaceuticals are not the subject of patents. Generic manufacturers like Cipla in India can manufacture anti-retrovirals and other pharmaceuticals without the need to apply for a compulsory license because pharmaceutical patent protection was not available in India when patent applications were filed in other countries. However, this is not the case in all developing countries. For example, it has been found that thirteen of the fifteen important antiretroviral drugs currently on the market are patented in South Africa.[4] Nevertheless, the need to have recourse to compulsory licensing has been avoided through voluntary licensing arrangements between patent holders and local manufacturers.[5]

It is likely that compulsory licensing activity will increase with both the implementation of TRIPS-compatible legislation in other developing and least-developed countries and the development of new and improved pharmaceuticals. However, the ability to manufacture under a compulsory license will only be of value if the *capacity* to manufacture exists.

COUNTRIES WITH INSUFFICIENT OR NO MANUFACTURING CAPACITIES

It probably goes without saying that all of the least developing countries have no capacity whatsoever to manufacture pharmaceuticals. The same can also probably be said for a number of developing countries. In such circumstances, the ability to issue a compulsory license enabling domestic manufacture will be of no practical assistance. This shortcoming is acknowledged in Paragraph 6 of the Doha health declaration, which provides:

> We recognize that WTO Members with insufficient or no manufacturing capacities in the pharmaceutical sector could face difficulties in making effective use of compulsory licensing under the TRIPS Agreement. We instruct the Council for

TRIPS to find an expeditious solution to this problem and to report to the General Council before the end of 2002.

There were extensive negotiations on this issue following Doha, culminating in the release of a draft decision by the chair of the TRIPS Council on December 16, 2002.[6] Yet despite the December 2002 deadline, resolution of this issue remained elusive until August 30, 2003, when the TRIPS Council finally agreed to adopt measures largely in accordance with the Chair's draft.[7] The key points of agreement and contention are discussed below.

Agreed Paragraph 6 Mechanisms

There are various mechanisms that may allow manufacture in one country (hereafter the supplying country) and importation into another country (hereafter the beneficiary country), even when a pharmaceutical is protected by patents in the supplying or beneficiary country or both. The TRIPS Council has agreed that the most appropriate mechanism for implementing Paragraph 6 requirements is through double compulsory licensing.[8] This means that the supplying country issues a compulsory license to a generic manufacturer to produce a certain quantity of pharmaceuticals for export to the beneficiary country, and the beneficiary country imports by way of a second compulsory license issued to a distributor. The difficulty with this mechanism is that TRIPS specifies that fairly stringent procedures must be followed for granting compulsory licenses. The decision by the TRIPS Council to adopt the double compulsory licensing option has been heavily criticized on this basis. It has been argued that this mechanism is unworkable and could actually be a step backward from the flexibility that previously existed in TRIPS.[9]

The requirement in Article 31(f) requiring that use must be *predominantly to supply domestic market* is the most significant limitation. The TRIPS Council decided that the appropriate way to deal with this problem is to formally declare that reliance on this provision should be waived whenever Paragraph 6 mechanisms are employed.[10]

The Main Point of Contention: Which Diseases Should Be Covered?

The question of which diseases can be covered by the Paragraph 6 mechanisms has remained the most contentious issue. On the one hand it could been argued that because Paragraph 1 specifically lists HIV/AIDS, tuberculosis (TB), malaria, and other epidemics, these should be the only diseases that are covered. The alternative viewpoint is that there should be no limits on the diseases that can be covered. This is supported by Paragraph 4 of the declaration, which expressly states that nothing in TRIPS prevents measures to protect public health. However, the pharmaceutical industry has expressed concern that this would allow generic producers "in perpetuity to supply copies of products for oncology, for mental health, for diabetes, even for lifestyle drugs."[11]

Médecins sans Frontières, one of the prominent nongovernmental organizations in this area, has argued that "paragraph 6 had *never* been intended to address only national emergency situations [emphasis added]."[12] However, if Paragraph 6 is read too expansively it could be seen as undermining the patent system by adjusting the balance too far in favor of access at the cost of incentive to innovate. It is fairly safe to assume that countries with well-established pharmaceutical industries would not agree to such a fundamental shift in balance. In order for the system to continue to be workable in its existing form, there has to be some limit on the applicability of Paragraph 6 mechanisms, perhaps by restricting them to public health crises.

If Paragraph 6 is restricted in this way the remaining issue is whether it can be used for public health crises other than HIV/AIDS, TB, malaria, and like epidemics. Arguably, cancer, genetic disease, parasitic infections, injuries caused by natural disasters, industrial accidents, or terrorism could all legitimately fall within the concept of public health crises, justifying the supply of medicines through Paragraph 6 mechanisms. This broad interpretation is justified on the basis that even if generic products to treat such diseases are manufactured under Paragraph 6 mechanisms, they are unlikely to have any real effect on the incentive to innovate, because the target markets for such products are not the same as the markets for patented products. After all, the primary justification for implementing such measures in the first place is that the patented products are not available in those markets because they are unaffordable.

Despite pressure from the United States to limit Paragraph 6 mechanisms to a fixed list of epidemics, ultimately the TRIPS Council agreed that there should be no such limitation. The TRIPS Council decision of August 30, 2003, defines pharmaceutical product as follows:

> any patented product, or product manufactured through a patented process, of the pharmaceutical sector needed to address the public health problems as recognized in paragraph 1 of the Declaration.[13]

OTHER ISSUES

Which Countries Are Beneficiary Countries?

The issue here is whether this provision should be interpreted broadly to include countries that are unable to manufacture a specific product, for whatever reason, or narrowly to include only those countries that lack any manufacturing capacity whatsoever. From early stages of the debate on Paragraph 6 mechanisms there was general agreement that all of the least-developed countries should automatically qualify as beneficiary countries and all developed countries should exclude themselves.[14] However, the contentious issue has been how to determine which other countries should be included. There are two main options. First, countries should be required to meet certain objective criteria. Second, countries should be able to assess *for themselves* whether they qualify. As to the first, there has been strong opposition to the use of preestablished lists based either on income or on human development criteria. Such opposition appears to be well-founded. For instance, the mere fact that a country does have some manufacturing capacity in the pharmaceutical sector does not necessarily mean that it has the capacity to manufacture the drugs needed for public health crises. A country may have the capacity to manufacture relatively simple drugs, but not the more complex antiretrovirals. The TRIPS Council agreement of August 30, 2003, presents a sensible compromise involving self-assessment based on objective criteria.[15]

Which Countries Are Supplying Countries?

The issue here is whether the supplying country has to be a developing country itself, or whether it can be a developed country. The advantage of allowing a developed country to generically manufacture under a compulsory license is that it increases the range of countries that could supply to beneficiary countries. It may also ensure superior quality and safety of the manufacturing process. Of course, it could be argued that this enables rich countries to subsidize their generic manufacturers to produce drugs for export to beneficiary countries at or below cost. However, the concern of some governments is that this increases the risk that pharmaceuticals produced for export to beneficiary countries will be diverted into the supplying country's own market or other rich-country markets.[16] A reasonable counterargument is that developed countries should put in place their own safeguards to ensure that diversion is prevented or kept to a minimum (as discussed later in this chapter). Appropriately, the TRIPS Council decision does not provide any limitations on which countries qualify to be supplying countries.[17]

Which Products Are Pharmaceutical Products?

The debate here is whether the Paragraph 6 mechanisms are limited to drugs in their final form, or whether they include active pharmaceutical substances and formulations used in the manufacture of drugs and manufacturing processes as well as other end products including diagnostics and vaccines. It has generally been accepted that intermediate products and processes should be included, but the inclusion of diagnostics and vaccines has been more contentious. There is strong justification for including them on the basis that they are as vital, if not more so, in dealing with public health crises as pharmaceuticals. It is only with proper diagnosis and prevention that epidemics will be brought under control. The TRIPS Council has accepted that diagnostic kits should be included.[18]

PRACTICAL PROBLEMS

In addition to these definitional problems, there are practical problems associated with the proposed Paragraph 6 mechanisms.

Generic Manufacturers

One of the essential requirements in achieving the goal of universal access to medicines through Paragraph 6 mechanisms is finding generic manufacturers that are willing to manufacture drugs under compulsory licenses for export purposes. Practically, there may be few generic manufacturers that are willing to take on this responsibility. It may take considerable time, training, and expense to provide the necessary infrastructure to produce a new drug and to satisfy safety and efficacy requirements. The economic feasibility of this is doubtful when demand is limited, for example, if supply is limited to a single beneficiary country. It may be necessary for supplying countries to provide subsidies or tax advantages in order to encourage manufacture for export.

Remuneration

Article 31(h) of TRIPS requires that the patent holder be paid adequate remuneration for manufacture under a compulsory license. Interestingly, this issue has been given little attention in discussions relating to Paragraph 6 mechanisms. The provision of adequate remuneration may become a major obstacle to the provision of affordable medicines if the payment of such remuneration necessitates a significant price increase. The task of TRIPS Council members was to find an appropriate solution that gives the patent holder due reward for use of the invention but does not impose too onerous a burden on the licensee, thereby ensuring that products produced under license are priced as cheaply as possible. The TRIPS Council agreement of August 30, 2003, provides only a partial solution to this issue. Paragraph 3 provides that only one remuneration payment is required to be paid pursuant to Article 31(h) of the TRIPS Agreement, even though strictly speaking two payments should be made because two licenses are issued.[19] This paragraph also states that in determining adequate remuneration, the economic value to the importing member should be taken into account. However, it remains to be seen exactly how the adequacy of remuneration is assessed, and this may prove to be the most significant limitation on the practical utility of the Paragraph 6 mechanisms.

Diversion

There is a risk that products manufactured under compulsory license will be diverted to developed-country markets. Additional safeguards have been agreed to concerning deterrence of diversion, including distinctive packaging, labeling, and coloring or shape of the product as well as controls on production and export in the supplying country and on distribution in the beneficiary country.[20] These safeguards could be provided in domestic legislation or in the terms of the compulsory licenses.

THE NEED FOR OTHER MEASURES

Whilst this approach to dealing with public health crises facing the developing world is undoubtedly thinking "inside the box,"[21] it does demonstrate an admirable resolve on the part of participants at TRIPS meetings in addressing this issue. Agreement on the utilization of compulsory licensing, particularly through Paragraph 6 mechanisms where there is a lack of manufacturing capacity, is an important step in the right direction. The agreed limitations on the use of compulsory licensing and safeguards against diversion should provide appropriate assurances that the existing balance will not swing too far in favor of access at the cost of incentive to innovate, irrespective of the nature of the disease in question.[22] Indeed, there may be no real effect on the existing balance because generic products produced under compulsory license will not enter the market for patented products.

From the perspective of the end user, medicines produced under compulsory license may still be unaffordable. Consequently, Paragraph 6 mechanisms are only likely to provide a solution if they are combined with other mechanisms. These might include funding of the purchase of medicines by international organizations, charitable organizations, and governments of developed countries; transfer of technology; improved health-care services; education; and ongoing biotechnological and related research into the cause of disease that focuses on prevention as well as treatment.

NOTES

This chapter is an abridged version of an earlier article by the author: D. Nicol, "Balancing Access to Pharmaceuticals with Patent Rights," *Monash Bioethics Review* 22 (2002): 50–62.

1. H. Binswanger, "HIV/AIDS Treatment for Millions," *Science* 292 (2001): 221–223.

2. Commission on Intellectual Property Rights, *Report: Integrating Intellectual Property Rights and Development Policy* (2002), 36, www.iprcommission.org/text/documents/final_report.htm (November 1, 2002).

3. J. Nielsen and D. Nicol, "Pharmaceuticals and Patents: The Conundrum of Access and Incentive," *Australian Intellectual Property Journal* 13 (2002): 21–40.

4. Commission on Intellectual Property Rights, 41.

5. Aspen Pharmacare, "Aspen Pharmacare Receive Voluntary License from GlaxoSmithKline on Anti-Retroviral Patents in SA," (2002), www.aspenpharmacare.co.za/showarticle.php?id=135 (November 1, 2002).

6. TRIPS Council Chair, *Draft Decision on Implementation of Paragraph 6 of the Doha Declaration on the TRIPS Agreement and Public Health* (2002), www.ictsd.org/ministerial/cancun/docs/TRIPs_para6_16-12-02.pdf (February 21, 2003).

7. TRIPS Council, *Implementation of Paragraph 6 of the Doha Declaration on the TRIPS Agreement and Public Health* (2003), www.wto.org/engish/tratop_e/trips_e/implem_para6_e.htm (September 26, 2003).

8. TRIPS Council, 2003.

9. C.P. Chandrasekhar and J. Ghosh, "WTO Drugs Deal: Does It Really Benefit Developing Countries?" *The Hindu Business Line* (September 9, 2003), www.thehindubusinessline.com/2003/09/09/stories/2003090900140900.htm (September 26, 2003).

10. TRIPS Council, 2003, paragraph 2.

11. "Phrma Looks to Limit Countries: Diseases Subject to New TRIPS Flexibility," *Inside US Trade* (October 18, 2002): 6.

12. International Centre for Trade and Sustainable Development, "Last Minute Attempt to Save TRIPS & Health Discussions," *Bridges Weekly Trade News Digest* (February 12, 2003), www.ictsd.org/weekly/03-02-13/story2.htm (February 21, 2003).

13. TRIPS Council, 2003, paragraph 1(a).

14. TRIPS Council Chair, paragraph 1.

15. TRIPS Council, 2003, paragraph 1(b) and annex.

16. International Centre for Trade and Sustainable Development, "WTO Stalemate Persists over Developing Country Imports of Generics,"*Bridges Weekly Trade News Digest* (October 24, 2002), www.ictsd.org/weekly/02-10-24/story3.htm (October 28, 2002).

17. TRIPS Council, 2003, paragraph 1(c).

18. TRIPS Council, 2003, paragraph 1(a).

19. TRIPS Council, 2003.

20. TRIPS Council, 2003, paragraph 2 (b)(ii).

21. S.R. Benatar, "Improving Global Health: The Need to Think 'Outside the Box,'" *Monash Bioethics Review* 22 (2003): 69–72.

22. D.B. Resni and K.A. DeVille, "Bioterrorism and Patent Rights: 'Compulsory Licensure' and the Case of Cipro," *The American Journal of Bioethics* 2 (2002): 29–39, and associated commentaries 40–52.

14

Media, Biotechnology, and Culture

Margaret Coffey

A global media is producing news of biotechnological advances that is often detached from analysis of the cultural context that produced them or that will exploit them. But media workers should be alert both to the assumptions that underlie bioethical discussions and to their own engagement with those assumptions. They must learn the histories, languages, and traditions—and the diversity—of their own communities as well as "the global community."

MOLLY NASH AND THE MEDIA

The globalization of ethics and the role of the media in this process were exemplified in the widespread media coverage in 2000 of the story of Molly Nash, a little girl whose U.S. parents had gone to extraordinary lengths to save her life—and for whom science had come up trumps. Molly suffered from a genetic disease called Fanconi's anemia where bone marrow fails progressively and blood cells are not produced, so that there is increasing debility, an impact on IQ, the development by the age of six or seven of leukemia, and, inevitably, death. Molly's parents decided to undergo in vitro fertilization (IVF), even though they were fertile, in order to achieve the implantation of an embryo both free of the disease and with tissue matching Molly's. Along with the aim of having a healthy child, their additional aim, and their pediatrician's, was to save Molly's life by acquiring material to enable a hematopoetic stem cell transplant. To achieve this aim, a preimplantation embryo biopsy had to include testing for the genetic abnormality associated with Fanconi's anemia as well as typing for human leukocyte antigen (HLA), in order to identify matching tissue that is standard for transplantation.

In 1996, the Nashes' first attempt at IVF produced an embryo that did meet with their criteria, being both disease negative and matched with Molly. However, the embryo transplantation failed. It took a total of five IVF attempts before the Nashes and their doctors effected a successful implantation. Molly was already six and showing signs of leukemia in August 2000 when Adam was born. The transplantation process began a couple of weeks after his birth, using the infant's umbilical cord blood. If the cord blood transplant had not been successful, the parents intimated that they would have then proceeded to use Adam's perfectly matched bone marrow.[1] (Bone marrow transplantation requires extracting bone marrow through the hip, and it carries risk to the donor and to the recipient.)

The steps taken by the Nash family in conjunction with Molly's treating doctors continue to be the subject of considerable discussion by experts in the fields of ethics, health, and law. However, discussions of this expert nature are framed quite differently, very often, from discussions conducted in the media. These more specialized contexts of expert discussion are not usually accessed by the general public, except insofar as the media seeks experts to proffer their opinions, or experts volunteer their points of view to the media. Expert discussions proceed in distinctive ways, employing elaborated vocabularies and paying particular attention to exactness of meaning. They can be continued at the next conference or seminar, reviewed in the next journal article, modified in the next book, in the confidence that there is certain continuity in the membership of the discussion. Expert discussion not only involves particular kinds of responsibilities and competencies, it also involves special safeguards, such as peer assessment and review. Different circumstances prevail in the media: it seems to me that it is important to understand the differences and the effect they may have on the framing in the media of ethical discussion. In my observation, some of these differences became evident in the purveying to the public of the Nash story and in its coverage by Australian media.

It was never in question that there would be intense media interest in the story of the Nashes and their recourse to vanguard medical science. What is intriguing is the degree to which this global media interest was channeled and directed by the public relations office of the University of Minnesota, the institution that is host to the scientists and medical practitioners who carried out the transplantation procedure.[2] An emblematic photo of Molly holding her baby brother appeared on front pages across the globe after a photo opportunity organized by the university. A free download of a "family photo" was made available on the university's website. The university's professor of pediatrics, John Wagner, who looked after the Nashes, said that the aim was to initiate a public discussion and "to be able to make a point that embryo research can have positive attributes and can be useful in helping save lives."

In a university press release, Professor Wagner claimed a comprehensive role for his institution:

> The University of Minnesota is committed to leading not only the science and medicine in areas like stem cell research, but the ethics and legal issues raised by the research as well. The University of Minnesota is uniquely positioned because it is home to a broad spectrum of academic disciplines, including the schools of medicine, law and public affairs, the Center for Bioethics and the Stem Cell Institute.[3]

The press release did not explain further what he meant by "leadership" in ethics and legal issues. At the same time, a political dimension was implied when Professor Wagner added, "we think that there should be a re-evaluation of [United States] government bans on such work [in the public sector]," so that it would not be forced into the private sector where "we were afraid that it could be more misused under the current regulations."[4]

The University of Minnesota's bioethicist, Dr. Jeffrey Kahn, had been involved in ethical appraisal of the Nash case and the policy questions it raised for quite some time, particularly during Mrs. Nash's pregnancy (and he contributes to this volume a detailed and insightful discussion of the ethical implications in the United States context of the Nash case). When the procedure was made public, his engagement in media forums contributed to how media framed the discussion. In one such media discussion, Dr. Kahn, who directs the University of Minnesota's Center for Bioethics, said that "...what it [the Nash case] raises is not so much an issue about whether the Nashes should have done what they did, but how that technique might be applied in different ways. I think the only really limiting factor in terms of what you would test for at the pre-implantation stage, is what genetic tests we have access to." He too, he averred, would do it for his child.[5]

This discussion of the Molly Nash story occurred in a broadcast on Australia's Radio National, where the host interviewed both Dr. Kahn and Professor Wagner. The program, "Genetic Technology and Ethics," was part of a highly esteemed and popular regular series, *The Health Report*, in which the host is himself a medical doctor as well as a broadcaster of distinction.[6] Kahn's declaration that he would consider using the means employed to save Molly Nash's life to save his own child's life came in response to the host's opening remark that, "Most parents, I suspect, would shrug their shoulders and say what's the problem? I'd do it for my child." When the host asked Dr. John Wagner, "Is there even an ethical issue? . . . Really the bottom line is that they're having a baby to help save another baby's life. Is there anything so wrong with that?" Dr. Wagner replied, "Well I don't believe so, and that's the reason we obviously allowed this to occur, and this is the reason why we suggested that it was a technology that was available.

But I think that that's not really the debate at issue. I think the debate at issue is—what next?" The ongoing consultations, discussions, and debates that had engaged Kahn, Wagner, and the Nash family throughout the case were thus elided. (I am informed by Dr. Jeffrey Kahn that on some issues he disagreed strongly with Dr. Wagner. Some of these issues may be inferred from the paper published in this volume, where, for example, Dr. Kahn draws attention to the risks to prospective donors of bone marrow and proposes third-party review to make sure appropriate risk–benefit balance exists when children are used as donors. In the Molly Nash case, a bone marrow transplant was proved unnecessary by a successful cord blood transplant.) Similarly, the academic independence enjoyed by Wagner and Kahn was displaced in the media by their identification with the institution, "the University of Minnesota," both in the press release announcing the Nash procedure and as they were contextualized in the radio program. The kind of contraction described in these examples is simply what happens in the media. So is the kind of personalization evidenced in Kahn's admission that he too "would do it for his child."

Radio programs, newspaper articles, television news stories, and indeed university public relations departments all work according to the requirements and conventions of media storytelling. The story is told to the moment, no matter how sophisticated the media outlet. The story must be clear and succinct, according to the specific stylistic demands of each format and outlet. Many formats and outlets allow little or no room for foregrounding the story. Characters are crucial—they ought to number no more than necessary—and character development is limited. Moreover, contributors to media stories—even more so, subjects of them—may have little or no influence over the use made of their material. They have no influence over the distribution or the consumption of their contribution. The Australia Broadcast Corporation (ABC) Radio National program about the Nash case referred to previously was broadcast throughout Australia on the country's only national radio station; as such a national broadcast it would have been the only widely available supplement to the brief news stories reporting the treatment of Molly Nash. In addition, the program was broadcast on Australia's external service, Radio Australia, to the countries of the Asia Pacific region. It was also broadcast around the world via the Internet and it remains available in transcript form on the ABC Radio National website. I am part of this program-making enterprise—I wear a pair of media lenses—but I think that it would be ingenuous to argue that this program did not have a unique significance, different in an exemplary way from the significance of a discussion that might occur on a campus during an ethics conference. It exemplified at least the difficulties inherent in conducting ethics discussions via the media. I have to confess though that it was some months after the broadcast that the provocation of the Salzburg Seminar "Biotechnology:

Legal, Ethical, and Social Issues" led me to reflect on the ideological dimensions of the program's discussion of the Molly Nash case.

In the aftermath of the Salzburg Seminar I thought about the notable symmetry of the discussion. The exchange was pragmatic, informed by the capacities of new technologies and an agreed-upon need to "draw lines" on a rational basis to their application, and it was detached from any explicit principle other than a desire to avoid "ratcheting up risk" in a context where it was difficult to restrict reproductive liberty.[7] According to Wagner, "we should at least develop a process by which to consider whether a given request is reasonable or not reasonable." And in Kahn's view it was advisable that there be some regulation—and funding—by a government agency, given the absence in the United States of either regulation of the private sector or public funding for embryo research. The broadcast did not provide a forum in which to query the University of Minnesota's leadership ambitions in the fields of law and ethics or the power made manifest in its involvement in the novel combination of strategies for Molly Nash's treatment.

AUSTRALIA, THE UNITED STATES, AND CULTURAL CONTEXTS

Now, Australia and the United States are both advanced Western democracies where standards of health care and education are high and where scientific research is proceeding apace. In this context, it is very easy to write articles and make radio and television programs that give the impression that we have achieved a level of refined abstraction in our approach to difficult bioethical issues, where our focus is not on people but on the person as a bundle of concepts—autonomy, free will, self-determination—abstracted from any particular culture or context. It is possible to imagine a continuing public conversation between Australians and Americans on bioethical issues, such as the one constructed by that radio program, informed by scientific opportunity on the one hand and a kind of ethical rationalism on the other. In this conversation, Australians and Americans would go on talking as if there were no differences between them. This is because, somehow or another, we have arrived at a conclusion about our common core rationality that suffices as a basis for conversation. We have arrived, we might imagine, at a "global ethics," which would allow us, shorn of the distractions of culture or place or time, to cool-headedly consider "the options."

Yet, in fact, there are many differences between us—between us and also amongst us. It may even be that the complexities of our respective societies are increasing, rather than diminishing. Not surprisingly, it is impossible to arrive at any notion of who Australians or Americans are without canvassing these differences in their whole historical, economic, political, racial, religious, and cultural range. Moreover, we understand ourselves in comparison not just with either the United States or Australia but also with the rest of the world.

Reflecting on this, it becomes a puzzle as to why a program—any program—on genetic technology and ethics would discuss the story of Molly Nash and the efforts to treat her disease without referring to the distinctive context or culture that produced those efforts. Why would the discussion—any discussion—not advert, for example, to the resources available to the Nash family who utilized expensive IVF techniques repeatedly, had access to benchmark research and medical techniques in an unregulated environment, and moreover had the institutional support of a university as they relayed their experience to the world public via the media? Indeed, is it not interesting sociologically to note the wide scope of the university's engagement, either as an institution or via its staff, in the Nash story? I wonder further about our alertness to the distinctive character of the conversation conducted in the media by the University of Minnesota via its representatives in Molly Nash's pediatrician and its bioethicist adviser? Do we in the media—at least in the United States and in Australia and perhaps in other Western countries as well—recognize in this kind of conversation the expression of an ideology, one that many of us who work in the media might share, for which secularization is normative, science is authoritative, progress is inevitable, and rationality is key?

Much media criticism answers questions of this kind pessimistically, in terms of the media deepening and extending the corrosive characteristics of modernity. Identity and connection (and, therefore, values) of the kind sustained by cultures and communities are put at risk by modernity, so the critiques go, and the media further destabilize identity and connection, substituting specious and self-serving versions.[8] Moreover, the media has a totalizing influence so that it also corrupts at the level of cultural quality, blunting the tools and habits of nuance, distinction, and discrimination. In short, criticism of the media places us as victims of "media manipulation and modernity's confines,"[9] our focus narrowed to the issues of modernity: notions of autonomy, the maximization of pleasure, the minimization of pain, the uses of technology, and the inadequacies of tradition and of the past. So, what about a media with a critical consciousness of its late modern context? Such a self-conscious media may be able to help us escape victimhood. The media could be a site for a rich, collective conversation that brings differences to light but without disregarding them. It could host a conversation based on an understanding of culture as an intrinsic part of human nature, which cannot be rationally dismissed.

The story of Molly Nash came initially to my attention via a daily broadsheet in my home city, Melbourne, Australia. That image of Molly and her infant brother supplied by the University of Minnesota took pride of place on page one. The accompanying facts were minimal: Molly had no biography other than that of disease and cure and no relationships other than those with parents, sibling, and doctors. It was not until the Salzburg Seminar "Biotechnology: Legal, Ethical, and Social Issues," in October 2001, that I heard any detailed account of the episode. And what remained with me most vividly from the lively discussion that followed was the silence in the seminar room of many of my fellow participants who came from countries where the priorities are food, water, hygiene, civil security, economic opportunity, and basic access to education. In a fundamental way the story of Molly Nash was irrelevant to their experiences. In conversations outside the seminar room, some made it clear to me that it was a story told in a "foreign" language. For example, an Indian delegate said:

> Our philosophy teaches us to approach the challenges you face in a calm way and take failures. We are trained to take failures much more easily. So a person who cannot afford this kind of technology even if he is aware that such a technology exists does not feel totally frustrated and dejected in life. If it is something beyond my reach I accept it.[10]

Another comment:

> You are destined to face this situation and, next, "Let me see how best I can do!" It is not sorrow for the rest of the life—of course it is sad—but he learns to put up with it and be happy within that situation, see what you can do to be happy and what you can do to make the child who is not like everyone else, how to make him happy. And the whole family, even the relatives, everyone contributes to that.[11]

A Pakistani delegate added, "In my country you can't talk about morals and ethics without considering religion and culture, obviously. I mean culture is in our country based on religion so you can't think about ethical issues or bioethics without thinking about your religion."[12] Yet in the context of our seminar's discussion of bioethics, religion made few appearances. When it did arise, it was invariably in a satiric remark from one or two Westerners who appeared to assume either hypocrisy or incoherence or simplemindedness when it came to religious approaches to these issues. The prevailing tone of the discussion was rather like that of the media coverage described previously: a tone abstracted from the influences on ordinary life in communities of tremendous diversity, long historicity, and profound socioeconomic and political strains. (The issues that did arise in the discussion included, for example, the risks associated with procedures the Nashes had signaled they would follow if cord blood transplantation did not succeed and the distinctiveness of the legislative environment in the United States.) Curiously, many people in this international gathering, including some Westerners, came from backgrounds of commitment to a religious identity: in their own persons, science, religion, and culture were in dialogue. But the dominant tone in our sessions seemed to preclude the development of a more inclusive and complex conversation.

CULTURAL ENGAGEMENT AND MEDIA RESPONSIBILITY

Mary Douglas has remarked on the parallel conversations often conducted by philosophers and anthropologists, the former about the rational foundation of ethics, the latter about the interaction between moral ideas and social institutions.[13] These conversations, she wrote, will never converge if each follows its present trajectory, to the detriment of any comprehensive, credible moral philosophy. Her parallel conversations mirror the conversation and the silence of the seminar room. And they mirror much of the discussion of bioethics in the media. This is curious because as media workers we do have unique opportunities to get to know communities, to learn the languages they speak, including the languages of religious traditions, and to acquire the vocabulary to broach questions of meaning and value.

Acquiring the right vocabulary is a crucial task. At one level, the task seems completed, since words like "dignity," "rights," "control," and "compassion" are already so freely invoked in media-based ethics discussion. However, we should not be satisfied that our use of these terms does their meanings justice. Dignity, for instance, sometimes seems to have more to do with being able to use a lavatory independently, or clean one's own teeth, than with anything inherent to the human being, especially the human being who has access to neither toilets nor toothbrushes. If these words are to have resonance as distinct from utility, they need the language of symbol and metaphor alongside them rather than the language of quantity and kind.

This is why it is important to engage religious and cultural traditions in the discussion.[14] Those values with which we begin our discussion are the products of those traditions. Thus we are surely better placed if we understand their long evolution—if, in other words, we learn our histories. Equipped with that understanding, we will then be less likely to take the easy route, epitomized in a remark by a frequent Australian contributor to these debates who described the Christian church as "clinging to ideas pertaining to a society of 2,000 or more years ago."[15] We will be alert to the principles of change and adaptation built into all religious traditions, and, to the benefit of our audiences, we will perhaps be able to read them at work within those traditions today. We should at least be able to avoid tags that are shorthand for our incomprehension of rich and dynamic sources of meaning for very many people. And we will begin to recognize in our own societies the way those traditions have contributed to the institutions and values we live with today.

Both the United States and Australia have the good fortune of a continuing encounter with indigenous peoples. Such encounter has compelled us to confront the deficiencies of problem-solving approaches that evade culture. In Australia, for example, evolving guidelines on research involving Aboriginal and Torres Strait Islander peoples recognize that these peoples form groups whose place within society is important to safeguard.[16] The guidelines both protect indigenous peoples and act as a vehicle for their more active involvement in the research process. The guidelines prompt collaboration that is at its best an exemplary form of cultural conversation where values are the bedrock of the exchanges and both parties learn a great deal about each other. Perhaps the most important point the guidelines make is that there truly are groups and societies where responsibilities are genuinely shared, as opposed to a kind of "pooled" individual autonomy and responsibility.[17]

So much media coverage of bioethical issues is built instead around the "hard case" or the individual story. The journalistic urge to place an issue within the context of an individual experience is surely understandable as a method of storytelling. However, it may be followed at the expense of diminishing the various interests of society in public debate. There is no image of society comparable, for example, to the image of Molly Nash holding her newborn brother. Perhaps it would take a cartoonist to represent "the suffering of society" or "the elation of society" in the face of individual autonomy-led decisions about biotechnology. Yet as media workers we need to keep in mind the power that images have of waylaying debate as well as focusing it. "Hard cases" may divert us from broader cultural themes, such as the relationship of the individual to society, the roles of the medical and scientific professions, the nature of the patient/doctor relationship, the diversity of our populations, and the appropriateness of law to the resolution of conflict. "Hard cases" may distract us from the difficulty of thinking and from the fact that sometimes problems cannot be solved.

Not that it is easy to find nuanced ways of engaging groups in any discussion or representing them in the media. Whether they are religious or cultural groups by definition, when they are marginalized or contained, as they often are, on the edge of our liberal, democratic, secular societies, they can appear seamless. Almost invariably this is a misperception. Whether a group is Buddhist, Christian, Islamic, Hispanic, Chinese, or Aboriginal, there is contest and adaptation going on. We need to be alert to this in the media so as to avoid cliché and also to enable us to invite the protagonists into the debate. Equally, as we look out from our own society toward other societies and traditions we should be wary of simplicities. For example, in the West we have come to think of a Chinese perspective on human life that entails a permissive attitude toward abortion and that therefore influences attitudes of Chinese people toward the Chinese government's birth control policies and indeed to eugenics. Research tells us, though, that in Chinese history (including in Buddhism and Confucianism), there have been different and opposing views about abortion. Furthermore, there is not one homogenous, culturally distinctive medical ethic that can be said to be "Chinese."[18] Instead, there continues to be debate, historical complexity, flux, and politics. In the West, in our respective cultures, circumstances are now teaching us this with respect to Islam. Maybe we need to understand it about ourselves.

The task of representing groups through the media brings difficulties similar to the task of representing society. Surveys of groups, or of society, cannot alone fulfill the challenge, even though in the media we respond to them with alacrity. Surveys are often devices to maintain media interest in specific issues, a strategy to which we seem sometimes oblivious in our reporting when we do not question their construction, commissioning, and usage. In any case, "popular opinion" is a problematic response to ethical issues, as Abraham Lincoln noted in 1854 when the Kansas–Nebraska Act put in place the notion that slavery was to be decided by "popular sovereignty." Lincoln understood that the act bypassed the "abstract truth" enshrined in the Declaration of Independence, that "all men are created equal," which was "applicable to all men and all times." The act put the doctrine of "popular sovereignty" in opposition to a fundamental moral—and constitutional—idea. To support it, Lincoln argued, risked blowing out "the moral lights around us."[19] Popular opinion may be problematic, but it is less so if it is alert and informed. In the face of these difficulties, media workers have critical responsibilities: to work to understand the traditions, languages, and histories of our communities, so that we can see the moral lights they bear in order to reflect these to readers, listeners, and viewers. Doing this may make genuinely fundamental criticism possible.

NOTES

1. "Genetic Technology and Ethics," *Health Report*, Radio National, Australian Broadcasting Corporation, March 19, 2001, www.abc.net.au/rn/talks/8.30/helthrpt/stories/s262813.htm.

2. University of Minnesota, "Umbilical Cord Blood Transplant Succeeds for Molly Nash," news release, January 4, 2001, www1.umn.edu/urelate/newsservice/news releases/01_01nash.html.

3. *Health Report.*

4. University of Minnesota news release, January 4, 2001.

5. *Health Report.* Many of these points were also canvassed in various issues of "Ethics Matters," a column contributed by Dr. Kahn to CNN.com. Dr. Kahn promoted and was also involved in nonmedia policy debate and discussion of the issues associated with the Nash case and stem cell research. For example, the University of Minnesota's Center for Bioethics hosted a forum on the ethics of preimplantation genetic diagnosis (PGD) three months after the birth of Adam Nash. The forum, planned prior to the birth, involved doctors, lawyers, and ethicists, none of whom thought that the Nash case represented an unacceptable use of PGD. The column is available at

www.ama-assn.org/sci-pubs/amnews/pick_01/prse0115.htm. Dr. Kahn chairs the University's Stem Cell Ethics Advisory Board.

6. Joli Jensen, *Redeeming Modernity: Contradictions in Media Criticism* (Newbury Park, CA: Sage Publications, 1990), 72.

7. *Health Report.* This is a point Dr. Kahn made also in the Winter 2001 edition of UMN's Center for Bioethics *Bioethics Examiner:* "We neither ask nor judge people's motives for having children, a longstanding and understandable policy in most liberal societies. So it will take significant justification, such as risk of serious harms to future children, to change the presumption from reproductive liberty." At www.cnn .com/2000/HEALTH/10/16/ethics.matters.

8. Jensen, 72.

9. Sociologist Alain Ehrenberg claims that private differences no longer undermine public space. Social recognition of these differences is the first principle of social action today and may be recognized in the content of television programming such as soap operas. However, recognition does not imply exploration, merely the naming of difference. See Alain Ehrenberg, *L'individu incertaine* (Paris: Calmann-Levy, 1995) and *Le culte de performance* (Paris: Calmann-Levy, 1991).

10. "Talking about Biotechnology," *Encounter*, Radio National, Australian Broadcasting Corporation, October 25, 2001, www.abc.net.au/rn/relig/enc/stories/s424006.htm.

11. *Encounter.*

12. *Encounter.*

13. Mary Douglas, "Morality and Culture," *Ethics* 93 (1983): 786.

14. The complexity opened by factoring culture into ethics discussion is interestingly considered by N. Yasemin Oguz in the Spring 2002 edition of the *Bioethics Examiner*, published by the University of Minnesota's Center for Bioethics.

15. Dr. Rodney Syme, *The Age*, May 17, 1995.

16. National Health and Medical Research Council, *Guidelines on Ethical Matters in Aboriginal and Torres Strait Islander Health Research* (Canberra: National Health and Medical Research Council, 1991).

17. Annette Baier, *Moral Prejudices: Essays on Ethics* (Cambridge, Massachusetts: Harvard University Press, 1994), 266.

18. Jing-Bao Nie, "Chinese Moral Perspectives on Abortion and Fetal Life: An Historical Account," *nzbioethics* 3, no. 3 (2002): 15–31.

19. William Lee Miller, *Lincoln's Virtues: An Ethical Biography* (New York: Knopf, 2003).

15

Tricksters, *The Plague*, and Mirrors: Biotechnology, Bioterrorism, and Justice

Katharine R. Meacham and Jo Ann T. Croom

The trickster (for example, the African spider "anansi" or the Native American coyote) both creates and confounds. Pandora plays the trickster role as the bringer of both knowledge that can unleash evil into the world and knowledge as the keeper of hope. The twenty-first century trickster is biotechnology. Biotechnology brings to our world both the power to create or cure, and the power to kill or destroy. The products and applications of genetic engineering include pharmaceuticals, recombinant bacteria for pollution remediation, gene therapy vehicles for treatment of genetic diseases, and new plant and animal strains for agriculture. However, that same knowledge allows for the genetic design of bioterrorist agents.

Garrett Hardin's famous essay, "The Tragedy of the Commons," identifies two kinds of problems: those that technology can solve and those for which there is no technological solution. Terrorism falls into the second category: despite many technological "fixes" that can be used to prevent, detect, and respond to various kinds of attacks, none are permanent solutions. Due to the potentially devastating effects of bioterrorism, serious attention is imperative both to preparation and response plans and to the more difficult, human phenomena of perspective, human nature, and ethical praxis.

This essay provides a snapshot of current empirical data related to the risks of bioterrorism and evaluates proposed responses. Its thesis echoes a line from Albert Camus' *The Plague*: "There's no question of heroism in all this. It's a matter of common decency. That's an idea which may make some people smile, but the only means of fighting a plague is—common decency."[1] Common decency takes many forms: from information and education to local, regional, and national preparedness; to international understanding and work for global justice. An ethic of common decency requires imagination, humility,

177

and vigilance. Camus' character Dr. Rieux warns against the presumption that a "war on terror" could be ever be "won." Instead,

> the tale he had to tell could not be one of a final victory. It could be only the record of what had had to be done, and what assuredly would have to be done again in the never ending fight against terror and its relentless onslaughts, despite their personal afflictions, by all who, while unable to be saints but refusing to bow down to pestilences, strive their utmost to be healers.[2]

BIOTERRORISM: DESCRIPTION, PREPARATION, AND RESPONSE

Imagine a bioterrorist event in Washington, D.C., with the release of enough smallpox virus to infect one hundred people (for example, the release of an aerosol in a crowded subway or a few smallpox "martyrs" in the early stages of disease, shopping in a crowded mall during the holiday season). For ten days, the victims are unaware of their plight, but on days eleven to thirteen a fever erupts with a violent headache, followed by a rash. If the victims travel to other cities such as New York or Tokyo during early disease, they spread virus particles via saliva from spots developing in the mouth. The red spots on the skin harden into pea-sized bumps that become reeking pustules. On the average, thirty of the original one hundred will develop more serious forms of the disease and die. Those who survive may be badly scarred or blind. Accurate diagnosis may not occur until the second wave of victims has been infected, and vaccinations may not begin until the third wave. A massive, coordinated public health response would be required to deliver and administer the vaccine needed to contain the epidemic around the world.[3]

Bioterrorism[4] is an efficient attack strategy because it does not rely upon conventional military divisions or armaments; it can succeed with minimal expenses and personnel. Biological weapons cost as little as 0.05 percent of conventional weapons and some can be produced for a $10,000 investment.[5] Contrary to the destruction caused by modern warfare, the infrastructure of the assaulted group is left intact.

While recent developments have spotlighted bioterrorism, it is not a new strategy. In 1346, the Mongols conquered the Black Sea port city of Kaffa, partly by hurling their own dead plague victims over the town walls. When the defenders fled to other parts of Europe, they may have sown the infectious agent (*yersinia pestis*) that began the epidemic of the Black Death.[6] Smallpox (*variola major*) played a pivotal role in the French and Indian war of 1754–1767, when the English deliberately gave infected blankets as agents of transmission to those Indian tribes loyal to the French. This technique continued to be used in the Western expansion.[7]

The two historically significant bioterrorist microbes—plague and small-pox—are on the Centers for Disease Control's (CDC's) current A-list of selected agents and threats.[8] Of the six agents listed, three (plague, anthrax, and tularemia) are bacterial; two (smallpox and hemorraghic fevers) are viral; and one is a natural toxin (botulism derived from the bacterium, *clostridium botulinum*). Sources for these bioagents vary. *Clostridium botulinum* and the anthrax organism, *bacillus anthracis*, are easily isolated from soils all over the world. The plague microbe, *yersinia pestis*, is endemic in certain species of wild rodent populations worldwide and can be transferred to individual humans by flea bite (bubonic form of plague) or, in an outbreak, can be transferred between humans via droplet infection (pneumonic form). Pneumonic plague is among one of the most infectious diseases known and can start from a single bacterium. Symptoms of systemic anthrax and pneumonic plague appear as early as twenty-four hours after infection and, to be effective, antibiotic therapy must begin promptly. If untreated, mortality approaches 100 percent.[9]

Emerging properties of modern life have made us more vulnerable than ever before to bioterrorism. There is now easy access to the genomes of many potential pathogens as well as to techniques and materials for isolating and cultivating microorganisms. The National Center for Biotechnology Information (NCBI) serves as a worldwide clearinghouse for current genetic analysis of organisms.[10] Its website lists the complete genome of all the following: the microbial agents (smallpox, plague, anthrax, and the bacterium that produces botulinum toxin) on the CDC A-list, as well as more than one hundred other bacterial species; a host of viruses, including severe acute respiratory syndrome (SARS); viroids, RNA-infective agents of plants; and plasmids, delivery packages for genetic engineering, for example, antibiotic resistance.[11] Such readily available information can be used for good or for ill.

With the appropriate genetic information and knowledge of relatively simple techniques in biotechnology, it is possible to produce designer rearrangements of genes to create super pathogens with increased resistance to antibiotics, enhanced infectiveness, and heightened virulence such as toxin production. Even advanced-placement high school biology students can carry out simple transformations of antibiotic-sensitive bacteria to antibiotic-resistant ones.[12] Underemployed scientists in disrupted economies have skills and perhaps incentive to perform biological engineering of more sinister types.

Not only do we now have access to Pandora's box of information, but we are also creating the ideal environment for epidemics. The increase in human population and development of industrialized societies produce large masses of people crowded together in cities where transmission of disease among persons is efficient. Health-care resources are already stretched beyond capacity and are often inadequate for current needs.[13]

Adequate immunization against common endemic diseases remains a challenge for many countries, and immunization against bioterrorist weapons is practically nonexistent for civilians.

International instability in political structures and economies increases the risk of bioterrorism. Ready-made pathogens from now inactive biological warfare laboratories in the former Soviet Union and South Africa tempt both potential terrorist buyers and sellers. It is reported that the United States has invested millions in retraining and employing former Soviet bioweapon scientists. Nevertheless, a relatively insecure facility in Kazakhstan still contains plague, anthrax, and smallpox.[14] Russian scientists are believed to have developed an antibiotic-resistant plague organism.[15] They are also reported to have refined both anthrax and smallpox for aerosol dispersal. One scientist who formerly worked with the Project Coast bioterrorist laboratory in South Africa attempted to sell to U.S. FBI agents strains of genetically altered *escherichia coli* fused with genes from the gas gangrene bacterium *clostridium perfringens*, which would cause severe illness. Bacteria that cause anthrax, plague, salmonella, and botulism were also available. While the proposed sale to the United States did not materialize, it is probable that there are other buyers eager to acquire these agents.[16]

Preparation is increasing worldwide for attacks using anthrax and smallpox.[17] As a bioweapon, inhalation anthrax has several advantages.[18] The spores are in a dormant stage that allows for easy transportation in the dry form preferred for aerosol distribution. If spores are not exposed to sunlight, they remain infective for more than one hundred years. Historically, mortality for untreated anthrax is 80 percent or higher. Prompt delivery of appropriate antibiotics may reduce the mortality rate; however, an antibiotic-resistant strain of anthrax was reported in 2003 by a research lab at Northern Arizona University.[19] The only disadvantage of anthrax as a bioweapon is that it is not easily communicable from person to person. Like anthrax, smallpox is stable and highly infectious in the aerosol form. Its morbidity is 30 percent, and it has an advantage of being highly communicable via saliva droplets.[20]

Estimates of damage that could be caused by deliberate use of these pathogens rely on mathematical models and on analysis of the results of accidental releases. One model postulates that under present conditions, the release of one kilogram of anthrax from a low-flying airplane or from the top of a tall building would result in the deaths of more than one hundred thousand people in a city the size of New York.[21] A 1979 anthrax outbreak, probably from a bioweapons plant in Sverdlovsk in the former Soviet Union, resulted in sixty-four deaths and thirty-two other illnesses over three months. Incubation periods ranged from two to ninety days. The official government explanation was that contaminated meat was the source. However, a 1994 study reported in *Science* argued that there may have been an accidental bioweapons release. The case reveals the importance of information to public-health and medical workers as well as citizens.[22] Models and accidental releases also reveal the need for adequate biosensors for early detection, trained and readily deployable public-health and emergency medical personnel, and sufficient stocks of antibiotics available for efficient distribution. One study recommended that antibiotics be distributed prior to an anthrax attack so that they are available within hours of exposure.[23] Local centers for distribution to individuals need to be established, with good access for all citizens.

Neither smallpox nor anthrax would be hazardous for immune populations. Yet, while vaccines for both agents exist, there have been problems with each. The anthrax vaccine is produced from a mixture of harmless proteins from the bacterium, but its efficacy in the prevention of inhalation anthrax is untested. Adverse reactions to the anthrax immunizations in the U.S. military (1998–2001) occurred in fewer than 1 percent of those immunized and were mild. Nevertheless, incomplete information and education led the public to distrust the program.[24] Research continues in an effort to find a more effective, less expensive vaccine.[25]

The modern vaccination program for smallpox was so successful that the World Health Organization (WHO) declared smallpox to be eradicated in 1980, three years after the last known naturally occurring case in Somalia.[26] With no smallpox outbreaks in the world since 1978, the risk of contracting the disease became nil—certainly much lower than the risks associated with the vaccine. Thus, the production of vaccines greatly diminished. With the renewed threat of smallpox as a bioterrorist tool, however, that risk–benefit profile is shifting. Medical authorities in the United States are seeking more careful and thorough analysis of responses to vaccination programs in controlled environments, for example, to public-health and military personnel, before vaccinations of the public.[27] Israel has a well-planned smallpox vaccination program, in which 15,000 first responders have been vaccinated and plans and stockpiles of vaccines are ready for the entire population in the event of an attack.[28]

Many medical professionals and first responders are reluctant to be vaccinated.[29] This is because of complications and contraindications associated with the vaccine. Complications include progressive blindness, encephalitis, and death (risk of one to three per million).[30] Contraindications for the vaccine occur with those who already have compromised immune systems (HIV-infected persons, newborns, chemotherapy patients), pregnant or nursing women, those with heart disease, eczema and certain other skin conditions, or anyone living or working with people who have those conditions.[31]

Ethical implications must be considered carefully by government officials, policy makers, and public-health workers. Since the risk of death, even if remote, is present, smallpox vaccinations should not be given unless the threat of bioterrorism is real and imminent. Who makes that call? Who should be vaccinated and in what order? What compensation exists for sick leaves and for costs of care if reactions to vaccinations occur? Should newly vaccinated workers be reassigned or furloughed so that they do not infect the increasing numbers of patients with compromised immune systems? Common decency demands that these questions be addressed before implementing vaccination programs. Individuals and officials need to understand the devastating effects of a smallpox epidemic so that they can make informed decisions.

What especially complicates the vaccination programs in high-target nations is the effect that the diversion of available resources to these bioterrorism prevention programs can have on the campaign for global immunization of children against potentially fatal common diseases such as measles, diptheria, tetanus, whooping cough, and polio. While great progress has been made in vaccination since the inception of the global campaign in 1974, it is estimated that 1,700,000 children die each year from vaccine-preventable diseases. The *Canadian Medical Association Journal* warns that "spending resources on improbable risks of biologic warfare posed by a rogue nation or extremist organizations at the expense of populations already gravely afflicted by holoendemic disease is irresponsible."[32]

Response to bioterrorist attacks that is effective and ethical clearly requires early detection, accurate data gathering, and the connectivity and efficiency of response networks. While high-tech devices for the detection of bioterrorist agents are being developed, it remains to be seen whether this will be a practical or affordable approach.[33] Basic information as to early symptoms of these diseases may not be part of the standard knowledge for some frontline medical workers.[34] One analyst claims that "the public health infrastructure is not merely an essential component; it may be the *only* component in the earliest phases" of response.[35] Moreover, it may be virtually nonexistent in some places around the world.[36]

The only treatments for smallpox are postexposure vaccination and antiviral agents. Experts recommend vaccination of anyone exposed to smallpox, regardless of other health conditions. If someone contracts smallpox, the risk of death is 200,000 times the risk from the vaccination.[37] Meanwhile, antiviral agents, if available, have undetermined efficacy.

All disasters begin locally, but appropriate response planning must take place at all levels. This evokes some critical questions. What is the balance of individual and communal rights with respect to isolation and quarantine, privacy, autonomy, and public health? How is the common good best achieved through education and sharing of information from governments and from local sources? How can there be the fair distribution worldwide of both sufficient information and supplies of medications and vaccines for prevention and treatment? What can be done about securing adequate facilities in hospitals and clinics for handling highly infectious diseases? How do all peoples have access to such care? Who pays for medications and care?

The potential for bioterrorism in agriculture is less known and potentially more destructive than bioterrorism against people. Disruption of the developed nations' food supplies by the deliberate introduction of plant or animal pathogens would be easy to achieve, hard to detect, and potentially devastating to the economy and the health of the citizenry. Many diseases of plants and animals are highly contagious, as demonstrated by the Malaysian swine paramyxovirus epidemic in 1999, which also killed one hundred people.[38] Although this epidemic was a natural phenomenon, the forced killing of more than a million pigs serves as a reminder of the vulnerability of food supply to disease and of humans to diseases that start in animal populations.[39]

It would be relatively simple to spread a disease throughout the human food supply for the following reasons. Modern agriculture has a limited selection of genetically pure strains for major crop species; agricultural crops are concentrated in relatively small areas (monocultures); and animal husbandry is highly concentrated into a few locations.[40] All that a potential terrorist needs is to locate an outbreak of the disease of choice (readily available on the Internet), gain access to a diseased plant or animal, collect a sample, and then inoculate a field of plants or a herd of uninfected animals.[41] Imagine the economic disaster that would occur if foot-and-mouth disease were deliberately introduced into the American beef system. Such an attack would be very difficult to prevent or trace, given the lack of detection sensors for plant and animal diseases, the openness of borders, and the problems in differentiating bioterrorist-initiated outbreaks from natural occurrences. As evidenced by the 2003 U.S. ban on beef from Canada with one diagnosed case of mad cow disease, the *report* of even a single infected animal can disrupt exports and negatively impact the economy of a nation.[42]

The need for the development of a comprehensive approach to biosecurity in agriculture has many parallels to human biosecurity. As with human biosecurity, there continues to be fragmentation in planning as well as a lack of coordination among agencies from the local to the international levels.[43] Coordination on the international level is particularly challenging, especially without full participation and mutually agreed upon enforcement of environmental and economic issues.

Not to identify the above dangers is to allow denial and fear to augment terror. There is a fourteenth-century tale of a dervish on the road outside of Baghdad who meets Plague.

"Where are you bound?" asks the dervish. "To Baghdad," responds Plague, "to kill a thousand." They chatted a while longer and then parted. Some months later, they chanced to meet again, and the dervish asked, "As I recall, you were going to kill a thousand. How is it that two thousand died?" "Ah, yes," replied Plague. "I killed a thousand. Fear killed the rest.[44]

An ethic of common decency will educate to increase knowledge and to reduce fear.

ETHICAL TECHNOLOGIES AND MIRRORS, IMAGINATION, AND JUSTICE

Bioterrorism prompts people to look with suspicion at others. That suspicion is in tension with the demands of ethical reflection. Ethical analysis requires honest self-examination, like that of looking into a mirror that gives us a sense to see ourselves from another's perspective. When individuals, nations, and policy-making bodies look into the mirror in the midst of preparing and planning for bioterrorist attacks, there is a critical initial decision that must be made: either everyone is a potential victim or everyone is a potential terrorist. Camus' character Tarrou frames it this way: "All I maintain is that on this earth there are pestilences and there are victims, and it's up to us, so far as possible, not to join forces with the pestilences."[45] This suggests a self-critical ethic that examines basic perspectives of both individuals and of nations. It requires a commitment to solidarity.

An attitude that everyone is a potential terrorist requires individuals to "circle the wagons," to guard against the "stranger." It is to demonstrate Freud's insight that "it is always possible to bind together a considerable number of people in love, so long as there are other people left over to receive the manifestations of their aggressiveness."[46] Rhetoric that characterizes nations, religions, and ethnic groups as "the enemy" feeds the attitude Tarrou abhors. It means joining the forces of the pestilences, by increasing fear, suspicion, and paranoia, and it consequently leads to more death and destruction.

In contrast, the assumption that everyone is a potential victim mandates an ethic of common decency that promotes concern for the common good, one form of which is social utility: the greatest good for the greatest number. In light of this concern for the common good, democratic societies that have tended to elevate individual rights, autonomy, and personal choice out of proportion to obligation to the common good face the challenge of mitigating of effects of bioterrorism. Authorities may recommend isolation, quarantine, and other limitations to individual autonomy out of deference to the public's health. There may be a reluctance to identify potential outbreaks due to fears of economic disruption and public reaction. Camus' medical and public-health authorities were also reluctant to identify the plague, leading Dr. Rieux to claim: "It's not a question of painting too black a picture. It's a question of taking precautions— [whether you call it 'plague' or not], we should not act as if there were no likelihood that half the population would be wiped out; for then it would be."[47] Responsible use of the media for educating the public about the most effective preparation and responses to bioterrorism can prepare people for appropriate protocols. In potential bioterrorist crises, the concern for the common good must be weighed against individual autonomy and freedom, otherwise the consequences are predictable and unconscionable. On the other hand, common decency is not honored when fears and threats are used to justify invasions of privacy that are not necessary for respectful, community-building preparation against bioterrorism.[48]

In a climate of increased fear of bioterrorism, scientists find themselves in a dilemma with regard to their work and this concern for the common good. One of the tenets of science is the sharing of information, a process vital for the generation of new knowledge. However, there now exists in the biological sciences a situation similar to that since the 1940s in the nuclear sciences— a potential for misuse of information in the generation of weaponry that threatens the stability and security of nations. One recommendation calls for the development of responsible stewardship, prescribed by international agreements, regulations, and professional codes.[49] In any case, curtailing biological research in an effort to prevent the potential development of bioterrorist agents risks diminishing the ability to develop biodefenses.

When we look deeply into the human mirror, we may recognize that the initial perspectival choice between seeing all people as either victims or terrorists is a false dichotomy.[50] Camus' character Tarrou warns:

> I know positively . . . that each of us has the plague within him; no one, no one on earth is free from it. And I know, too, that we must keep endless watch on ourselves lest in a careless moment we breathe in somebody's face and fasten the infection on him. What's natural is the microbe. All the rest—health, integrity, purity (if you like)—is a product of the human will, of a vigilance that must never falter.[51]

Tarrou's insight is that none of us is pure. Each of us has the capacity to "infect" others with hatred, fear, distrust, arrogance, and moral blindness. Nations, ethnic groups, and religions have this dual capacity as well. Vigilance is required.

Vigilance leaves no room for uninvolved spectators, whether individuals, groups, or nations. In Camus' novel, the journalist Rambert, finding himself trapped inside the town after it is quarantined, bribes officials in order to escape. However, once he decides to stop being an outside observer and joins the sanitation teams, he changes his mind. He still loves his fiancée; he still misses his home; but he no longer feels separate and apart. He says, "Until now I always felt a stranger in this town, and that I'd no concern with you people. But now that I've seen what I have seen, I know that I belong here whether I want it or not. This business is everybody's business."[52] Reflective action produces empathy.

The ethical move from concern with individual well-being to the realization that "this business is everybody's business" occurs through empathetic participation in the lives of the victims and involvement in work for the common good. Even in plague-infested fourteenth-century Damascus, Muslims, Jews, and Christians processed and prayed together in a spirit that "this business is everybody's business."[53] It can also occur through self-critical reflection in one's mirror that sees our commonality even with the terrorists. This perspective of solidarity with all peoples requires honesty and humility.

The realization that "this business is everybody's business," that everyone is a potential terrorist as well as a potential victim, calls for justice. Justice is like the proverbial elephant, surrounded by six blind men, each of whom defines the elephant according to where he stands and how he feels.[54] Many wars throughout human history have been and continue to be justified through retribution for perceived wrongs, and war crimes tribunals as internationally accepted mechanisms of retributive justice have proven their worth. However, they are not enough. Ironically, justice as retribution is also a rationale from the perspective of terrorists, who are angry over injustices engendered by the globalization of the economy and culture.[55] It takes no imagination to see that, as violence begets violence, retribution begets retribution. It is precisely this notion of "justice as retribution" that Tarrou decries as "joining forces with the pestilences." Moral imagination is needed.

There are two theories of justice that are clearly more imaginative than justice as simple-minded retribution. One is the theory proposed by philosopher John Rawls; the other is the theory of restorative justice. Rawls argues that justice is best seen as fairness. This fairness entails distributing society's "goods" in ways so that the least advantaged always stand to gain and are not preyed upon by the most advantaged. Although individual autonomy is not sacrificed for the common good, the common good is achieved through imaginative identification with the least advantaged. Restorative justice recognizes the harms that are done. It identifies the perpetrators of these harms and holds them accountable. Furthermore, it involves citizens in naming reparations and redress and then works for the restoration of the humanity of the offenders as well as reconciliation with the offended. The aim in restorative justice is to apply justice in order to heal and to restore human community.[56]

Just as bioterrorism is not new, neither is restorative justice a new idea. Its seeds are found in all the major religious traditions. Its theories find expression in the African philosophy of *Ubuntu* and the Christian "Just Peacemaking" movements.[57] The truth and reconciliation commissions of South Africa and of Liberia are examples of restorative justice in practice. Both restorative justice and justice as fairness can inform efforts to prevent terrorism and to help heal communities that are affected, even as we recognize that, plague-infested as we are, there may be no cure.

Not to join forces with the pestilences means that preparation for bioterrorist attacks should be rational, thorough, realistic, and fair. It means that response plans must recognize that when people know each other across religious, national, and ideological lines, then a common good can be imagined. It means working actively against outbreaks, with persistence and endurance. Camus suggests that even the plague can serve as trickster—tormenting us, for sure, but also revealing wisdom. At the end of the novel, Rieux keeps this in mind:

> that the bacillus never dies or disappears for good; that it can lie dormant for years and years in furniture and linen-chests; that it bides its time in bedrooms, cellars, trunks, and book-shelves; and that perhaps the day would come when, for the bane and enlightening of men, it would rouse up its rats again and send them to die in a happy city.[58]

We are left with an ambiguous and tragic-filled existence that challenges us to act for the good of the entire world, with fairness and common decency, even in the midst of overwhelming odds. Camus' narrator, who works stubbornly as a healer, admits that he wrote,

> so that he should not be one of those who hold their peace but should bear witness in favor of those plague-stricken people; so that some memorial of the injustice and outrage done them might endure; and to state quite simply what we learn in time of pestilence: that there are more things to admire in [human beings] than to despise.[59]

NOTES

1. Albert Camus, *The Plague* (New York: Vintage Books, 1972), 154.

2. Camus, 287.

3. Inspired from Richard Preston, "The Demon in the Freezer," in *Best American Science and Nature Writing* 2000, ed. David Quammen (New York: Houghton Mifflin, 2000), 153–178; Jennifer Lee Carrell, *The Speckled Monster,* excerpted in *Esquire* (June 2003): 30.

4. The Arizona Department of Health Services defines bioterrorism as "the use or threatened use, of biological agents to promote or spread fear or intimidation upon an individual, a specific group, or the population as a whole for religious, political, ideological, financial, or personal purposes." Arizona Department of Health Services, 2000, modified March 30, 2001; at www.hs.state.az.us/phs/edc/edrp/es/bthistor1.htm.

5. Jeffrey Pommerville, "A Primer on Bioterrorism," *The American Biology Teacher* 64, no.9 (November–December 2002): 649–657.

6. Steven M. Block, "The Growing Threat of Biological Weapons," *American Scientist* (January–February 2001), www.americanscientist.org/template/AssetDetail/assetid/14284?fulltext=true (accessed May 14, 2003). The term "Black Death" has been used for plague since the fourteenth century, because it refers to the bodies that turn black because of the accumulation of blood under the skin of victims.

7. William H. McNeill, *Plagues and Peoples* (New York: Anchor Books, 1976); U.S. Army, *Blue Book,* www.usamriid.army.mil/education/bluebook.html, 2.

8. "Terrorism and Public Health," CDC Public Health Emergency Preparedness and Response Site, www.bt.cdc.gov (accessed May 14, 2003).

9. Larry McKane and Judy Kandel, *Microbiology: Essentials of Application,* 2nd ed. (New York: McGraw-Hill, 1996), 654–655.

10. Among contributers to this site are the Chinese National Human Genome Center at Shanghai; the universities of Wuerzburg and Uppsala, as well as many in Europe and the United States; and private corporations.

11. National Center for Biotechnology Information website www.ncbi.nlm.nih.gov/.

12. Colony Transformation Kit from Carolina Biological Supply Company, http://www.carolina.com.

13. Laurie Garrett, *Betrayal of Trust: The Collapse of Global Public Health* (New York: Hyperion, 2000).

14. Andrew W. Artenstein, "Bioterrorism: Reflections on Anthrax and Preparations for Smallpox," *Medscape,* http://www.medscape.com/viewarticle/444476 (accessed May 19, 2003); John Bohannon, "From Bioweapons Backwater to Main Attraction," *Science* 300 (April 18, 2003): 414–415.

15. I. V. Domaradskij, *Biowarrior: Inside the Soviet/Russian Biological War Machine* (New York: Prometheus Books, 2003), in Christiane Amanpour, *Deadly Germs at Home in Russia: A 60 Minutes Special Report,* CBS, May 11, 2003, www.wwjtv.com/rooney/sixtyminutes_story_131201234.html.

16. Joby Warrick and John Mintz, "Lethal Legacy: Bioweapons for Sale, U.S. Declined South African Scientist's Offer on Man-Made Pathogens," *Washington Post,* April 20, 2003, www.washingtonpost.com/wp-dyn/articles/A58454-2003Apr19.html (accessed May 14, 2003); Joby Warrick, "Biotoxins Fall into Private Hands: Global Risk Seen in S. African Poisons," *Washington Post,* April 21, 2003, A01, www.washingtonpost.com/ac2/wp-dyn/A64518-2003Apr20 (accessed April 23, 2003).

17. Bohannon, "Bioweapons," *Science,* Report on the 5th International Conference on Anthrax in Nice, France, 2003.

18. Block; inhalation of about 10,000 spores needed.

19. Block. This strain was resistant to ciprofloxacin, the antibiotic of choice for treatment.

20. Mark Alpert, "Spotty Defense," *Scientific American* (May 20, 2003): 20–23.

21. Model developed by Lawrence Wein of Stanford University Graduate School of Business. *Proceedings of the National Academy of Sciences* (March 17, 2003), as reported in "Anthrax Threat Needs Aggressive Government Action Plan, Say Researchers," in *EurekAlert* (March 17, 2003), http://www.erurkalert.org/pub_releases/2003-03/su-atn031303.php (accessed March 13, 2003); also in Lila Guterman, "Death Toll in Airborne Anthrax Attack Could Exceed 100,000, Mathematical Model Finds," *Chronicle of Higher Education Daily News,* March 18, 2003, chronicle.com/cgi2-bin/prin...com/daily/2003/03/2003031801n.htm.

22. Meselson, Guillemin, Hugh-Jones, Langmuir, Popova, Shelokov, and Yampolskaya, "The Sverdlovsk Anthrax Outbreak of 1979," *Science* 266, no. 5188 (November. 18, 1994): 1202-8; at www.pubmed.com, PMID: 7973702, (accessed June 1, 2003); "Case Study: Sverdlovsk Anthrax Outbreak of 1979" at www.nbc-med.org, (accessed June 1, 2003).

23. *EurekAlert* (March 17, 2003); research team headed by Lawrence Wein of Stanford University.

24. The protocol for vaccination includes multiple injections over an eighteen-month period and annual boosters are required; the only systematic immunization that has been carried out is the inoculation of 500,000 in the U.S. military between 1998 and 2001. Data complied from this group indicates that adverse reactions occurred in fewer than 1 percent and were generally mild in nature. However, public perception of the safety of the vaccine remains tainted from adverse publicity during the first Gulf War.

25. Amanpour, "Deadly Germs," *60 Minutes.*

26. The original smallpox vaccine contained live smallpox virus from infected victims, but the modern vaccine is a derivative of the cowpox virus.

27. Brian Strom, "Review of the Centers for Disease Control and Prevention's Smallpox Vaccination Program Implementation: Letter Report No. 3" (May 23, 2003), www.nap.cdn/openbook/N1000508/html (May 29, 2003).

28. Sam Jaffe, "Smallpox Vaccination Tips from Israel," *BusinessWeekonline*, December 20, 2002, www.businessweek.com:/pting/bwdaily/dnflash/dec2002/nf20023220_6319.htm?gb (accessed June 3, 2003).

29. With current threats of bioterrorism, plans were made to vaccinate 500,000 front-line medical workers in the United States for smallpox, but the process has gone much more slowly than expected. As of March 21, 2003, fewer than 26,000 individuals had been vaccinated and some rare but troubling incidence of heart inflammation had been noted. Refer to "Smallpox Vaccination and Heart Problems," CDC Public Health Emergency Preparedness and Response, www.bt.cdc.gov/agent/smallpox/vaccination/heartproblems.asp.

30. Thomas Mack, "A Different View of Smallpox and Vaccination," *New England Journal of Medicine* 348, no. 5 (January 30, 2003), content.nejm.org/cgi/content/full/348/5/460 (accessed June 3, 2003).

31. Artenstein.

32. Editorial, "Global Vaccination Meets Global Terror," *Canadian Medical Association Journal* 167, no.8 (2002): 837. Since the inception of the global immunization program against potentially fatal childhood diseases, the incidence has dropped from 75 percent to 25 percent. Children in industrialized nations and where socialized medicine exists have higher percentages of vaccinations against childhood diseases than in, for example, Central African Republic and Haiti. "WHO Vaccine Preventable Diseases Monitoring System," *2002 Global Summary: Country Profile Selection Centre,* www1/who/int/vaccines/globalsummary/Immunization/CountryProfileSelect.cfm

33. Stephen S. Morse, "The Vigilance Defense," *Scientific American* (October 2002): 88–89; Rocco Casagrande, "Technology Against Terror," *Scientific American* (October 2002): 83–87.

34. Julie A. Pavlin, "Epidemiology of Bioterrorism," *Emerging Infectious Diseases* 5, no. 4, www.cdc.gov/ncidod/EID/vol5no4/pavlin.htm (accessed May 14, 2003).

35. Morse.

36. Garrett. Martha Salyers, M.D., M.P.H., Public Health, Regional Surveillance Team for Region 6 of the North Carolina Public Health System also contributed to this [our] chapter with her careful critical reading and suggestions of research and resources; her editorial skills; and her significant knowledge of bioterrorism preparedness, public health, and ethics.

37. Alpert; Carrell, *The Speckled Monster,* excerpted in *Esquire,* 30: "Odds of dying from smallpox during an epidemic: one in three. Odds of dying from . . . inoculation with live smallpox virus: one in 50. Odds of dying from modern vaccination: one or two in a million."

38. W. Wayt Gibbs, "Trailing a Virus," *Scientific American* (August 1999): 81–87.

39. In January 2003, Michigan had an "outbreak" of nicotine poisoning from intentionally contaminated ground beef; www.cdc.gov/mmwr/preview/mmwrhtml/mm5218a3.htm.

40. Mark Wheelis, Rocco Casagrande, and Laurence V. Madden, "Biological Attack on Agriculture," *Bioscience* 52, no. 7 (July 2002): 569–576.

41. Rocco Casagrande, "Biological Warfare Targeted at Livestock," *Bioscience* 52, no. 7 (July 2002): 577–582.

42. Mad cow disease is BSE: bovine spongiform encephalopathy.

43. Laura A. Meyerson and Jamie K. Reaser, "Biosecurity: Moving toward a Comprehensive Approach," *Bioscience* 52, no. 7 (July 2002): 598.

44. I. Edward Alcamo, *Fundamentals of Microbiology,* 6th ed. (Sudbury, MA: Jones and Bartlett Publishers, 2001), inside back cover, A-10.

45. Camus, 236.

46. Sigmund Freud, *Civilization and Its Discontents,* trans. James Strachey (New York: Norton, 1961), 61.

47. Camus, 47–49.

48. Some fear that the U.S. Patriot Act, for example, is being used as an excuse to increase surveillance, to decrease respect for privacy in the name of the common good, but out of an attitude that assumes everyone is a potential terrorist. Amnesty International's 2003 Report raises questions about eight countries (Liberia, Philippines, India, Russia, Israel, Palestinians, Nepal, Colombia, and the United States) and possible human rights violations in the name of "counterterrorism" actions; web.amnesty.org/report2003/focus2002_1-eng.

49. Gigi Kwik, Joe Fitzgerald, Thomas V. Inglesby, and Tara O'Toole, "Biosecurity: Responsible Stewardship of Bioscience in an Age of Catastrophic Terrorism," 42nd Interscience Conference on Antimicrobial Agents and Chemotherapy, April 24, 2003, on *Medscape* www.medscape.com/viewarticle/452338 (accessed May 14, 2003).

50. Research by the Josephson Institute of Ethics in 1989 revealed that most Americans (60–80 percent) think that most people who know them well would list them as one of the most ethical people they know. At the same time, most people (65–80 percent) think that their own ethics are "higher" than others! Josephson Institute of Ethics, "Examine Your Ethical Attitudes," in *How Do Others Make Moral Decisions?* (Westport, CT: Greenhaven Press, 1993), 200.

51. Camus, 235–236.

52. Camus, 194.

53. H.A.R. Gills, trans. and comp., *Ibn Battuta, Travels in Asia and Africa, 1325-1354* (New York: August M. Kelley, 1969), 305, 68–69. Fourteenth-century documents from Islamic and Christian sources indicate a range of responses to plague. Explanations ranged from the scientific (especially Islamic sources) to the theological (both Muslims and Christians).

54. Karen Lebacz, *Six Theories of Justice* (Minneapolis, MN: Augsburg, 1986), 9; David Roth, singer-songwriter, whose version of this story concludes with the line, "What you see depends on where you stand and how you feel."

55. Benjamin Schwartz and Christopher Layne, "A New Grand Strategy," *Atlantic Monthly* (January 2002), www.theatlantic.com/issues/2002/01/schwarzlayne.htm (accessed June 3, 2003).

56. Shereen Benzvy Miller and Mark Schacter, "From Restorative Justice to Restorative Governance," *Canadian Journal of Criminology* 42, no. 3 (July 2000), web8epnet .com/delivery.asp?tb=1&_ua=shn+56+4BC0&_ug=dbs+0+ln+en-us+sid+2 (accessed May 13, 2003).

57. *Ubuntu* means "humanity" but more: "a person is a person through other persons," as the Xhosa proverb defines it; in Tom Hewitt, "A Question of Justice," *Peace Review* 14, no. 4 (2002): 447–453; "Just Peacemaking" theory is usually credited to Christian ethicists Glenn Stassen and Walter Wink. The Rwandan term for restorative justice is *gacaca*, meaning "justice in the grass"; such a process first used a war crimes retributive model in Rwanda but also has worked to bring reconciliation across tribal lines; in *Christian Century* (August 28–September 10, 2002): 7.

58. Camus, 287.

59. Camus, 286–287.

Index

About the Contributors

THE EDITOR

Michael C. Branningan is executive director of the Institute for Cross-Cultural Ethics as well as professor and chair of the philosophy department at La Roche College. Along with numerous scholarly articles on ethics, biomedicine, and Asian thought, his books include *Ethics Across Cultures, Healthcare Ethics in a Diverse Society, and Ethical Issues in Human Cloning.* Born in Japan and raised in the United States, he received his Ph.D. in philosophy and M.A. in religious studies from the University of Leuven, Belgium. He has lectured widely on ethics and cross-cultural studies, and has received various national and international awards. His other interests include athletics, music, and the martial arts. He lives in Gibsonia, Pennsylvania with his wife, Brooke.

THE CONTRIBUTORS

Don Chalmers, Ph.D., is dean of the Faculty of Law, University of Tasmania, and chair of the Commonwealth Gene Technology Ethics Committee.

David Kum-Wah Chan, Ph.D., is assistant professor in the Department of Philosophy, University of Wisconsin, Stevens Point.

Margaret Coffey is broadcaster and producer for the Australian Broadcasting Corporation, Melbourne, Australia.

Jo Ann T. Croom, Ph.D., is professor in the Department of Biology, Mars Hill College.

Mylène Deschênes, JD, is research associate and projects director in the Centre de recherche en droit public, University of Montreal.

Heinrich Ganthaler, Ph.D., is associate professor in the Department of Philosophy, University of Salzburg.

Yuri M. Gariev, Russian Academy of Sciences, Moscow.

Stella Gonzalez-Arnal, Ph.D., is tutor in the Department of Gender Studies and Department of Philosophy, University of Hull.

Ryuichi Ida, Ph.D., is professor at Kyoto University Graduate School of Law and former chair of the International Bioethics Committee, UNESCO.

Jeffrey P. Kahn, Ph.D., M.P.H., is professor of medicine and director of the Center for Bioethics, University of Minnesota.

Martin O. Makinde, Ph.D., is professor and dean of the Department of Animal Science, University of Venda for Science and Technology.

Anna C. Mastroianni, JD, M.P.H., is assistant professor in the School of Law and Institute for Public Health Genetics, University of Washington.

Katharine R. Meacham, Ph.D., is chair of the Department of Philosophy, Mars Hill College.

Bushra Mirza, Ph.D., is assistant professor in the Department of Biological Sciences, Quaid-I-Azam University.

Michael J. Morgan, Ph.D., is chief executive of the Wellcome Trust, Hinxton, Cambridgeshire, United Kingdom.

Dianne Nicol, Ph.D., is lecturer at the Law School, University of Tasmania.

Edward Reichman, MD, Department of Epidemiology and Population Health, Albert Einstein College of Medicine, Montefiore Medical Center, Bronx, New York.

Susan E. Wallace, Department of Law, University of Sheffield.

Larissa P. Zhiganova, Ph.D., is senior researcher, Agriculture Studies Sections, Russian Academy of Sciences, Moscow.